Methods in Molecular Biology™

For further volumes:
http://www.springer.com/series/7651

Rational Drug Design

Methods and Protocols

Edited by

Yi Zheng

Division of Experimental Hematology and Cancer Biology,
Cincinnati Children's Hospital Medical Center, University of Cincinnati, Cincinnati, OH, USA

Humana Press

Editor
Yi Zheng
Division of Experimental Hematology and Cancer Biology
Cincinnati Children's Hospital Medical Center
University of Cincinnati
Cincinnati, OH, USA

ISSN 1064-3745 ISSN 1940-6029 (electronic)
ISBN 978-1-62703-007-6 ISBN 978-1-62703-008-3 (eBook)
DOI 10.1007/978-1-62703-008-3
Springer New York Heidelberg Dordrecht London

Library of Congress Control Number: 2012943829

Printed on acid-free paper

Humana Press is a brand of Springer
Springer is part of Springer Science+Business Media (www.springer.com)

Preface

Tremendous progress made in the past three decades of molecular biology studies of the cell signaling network has accumulated vast resources and knowledge that are ripe for harvesting for translational and therapeutic development. It is now clear that the traditional pharmaceutical Research & Development model may not be sustainable as evidenced by the recent flagging productivity in the pharmaceutical industry. New technological innovations in research of lead design, discovery, optimization, and validation, initiated by original ideas and concepts, may help mitigate the situation in the long run.

This volume of MiMM is aimed at offering some selected examples of novel methodologies involved in the dynamic and ever-changing field of rational drug design. The readers of this methodology reference book may adopt and modify the protocols and methods described in this volume in their efforts to translate unique bench-side ideas into discovering new leads of novel or established molecular targets in human diseases.

The areas covered in this volume include the following:

1. Virtual screening of chemical hits
2. Rational lead discovery by high-throughput screening
3. Combinatorial and fragment-based lead generation
4. Peptide-based drug discovery
5. Specificity and resistance
6. Effective delivery by RNA-based approach and nanotechnology
7. Animal models of lead validation

Structure-based virtual screening has emerged as an important tool in our quest to access novel drug-like compounds. There are a wide range of comparable and contrasting methodologies available in screening structural databases for hit identification. In Chap. 1 and several subsequent chapters, a strategy and several case studies utilizing the available virtual screening methods to identify hits and leads for small GTPases, phosphotases, or wnt pathway targets are presented. In addition to the traditional enzymatic targeting sites, such as the ATP-binding kinase domain or phosphotase domain, recent understanding of the structure–function relationship of many signaling proteins has allowed clever design of small molecule or peptidomimic inhibitors or activators against allosteric sites of the targets that significantly increase the specificity and efficacy. Examples of these also are discussed in the following chapters.

Complementary to the structure-based hit identification effort, mechanism-based screening approaches have proved fruitful leading to new leads of specific kinases and other enzymatic substrates. Examples in the successful application of yeast two-hybrid system, peptide aptamers, and peptide ligands are illustrated.

With the success of targeted therapy, development of drug resistance in disease cells has become an inevitable side effect. Recent progress in the drug discovery studies tackling the mechanism of drug resistance and applying rational designed second generation drugs to overcome target expression/mutation changes for several kinases pioneers this future direction of rational drug design. This area of studies is represented by the identification, characterization, and rational design of new inhibitors in BCR-ABL targeted therapy.

Finally, a recent example of designing antisense microRNA and an example of delivering membrane-impermeable molecules into cells by nanotechnology, which can expand potential therapeutic options of rational designed drugs, are presented.

Overall, these complementary approaches to identify and validate hits specifically targeting a biologically implicated target, to overcome ensuing drug resistance and specificity issues, and to improve delivery efficiency to the target cells should be useful for the readers to appreciate the fast pace in rational drug design that may benefit future drug discovery effort.

Cincinnati, OH, USA *Yi Zheng, Ph.D.*

Contents

Preface v
Contributors ix

1 On Setting Up and Assessing Docking Simulations for Virtual Screening 1
Jacek Biesiada, Aleksey Porollo, and Jaroslaw Meller

2 Virtual Ligand Screening Combined with NMR to Identify Dvl PDZ Domain Inhibitors Targeting the Wnt Signaling 17
Jufang Shan and Jie J. Zheng

3 Rational Design of Rho GTPase-Targeting Inhibitors 29
Xun Shang and Yi Zheng

4 Rational Design of Peptide Ligands Against a Glycolipid by NMR Studies 39
Wenyong Tong, Tara Sprules, Kalle Gehring, and H. Uri Saragovi

5 A Combinatorial Strategy for the Acquisition of Potent and Specific Protein Tyrosine Phosphatase Inhibitors 53
Sheng Zhang, Lan Chen, David S. Lawrence, and Zhong-Yin Zhang

6 Identification of Allosteric Inhibitors of p21-Activated Kinase 67
Julien Viaud and Jeffrey R. Peterson

7 Using a Modified Yeast Two-Hybrid System to Screen for Chemical GEF Inhibitors 81
Anne Blangy and Philippe Fort

8 Random Mutagenesis of Peptide Aptamers as an Optimization Strategy for Inhibitor Screening 97
Nathalie Bouquier, Sylvie Fromont, Anne Debant, and Susanne Schmidt

9 A Screening Strategy for Trapping the Inactive Conformer of a Dimeric Enzyme with a Small Molecule Inhibitor 119
Charles S. Craik and Tina Shahian

10 Use of a Fluorescent ATP Analog to Probe the Allosteric Conformational Change in the Active Site of the Protein Kinase PDK1 133
Valerie Hindie, Laura A. Lopez-Garcia, and Ricardo M. Biondi

11 Affinity Purification of Protein Kinases that Adopt a Specific Inactive Conformation 143
Pratistha Ranjitkar and Dustin J. Maly

12 Determination of the Kinetics and Thermodynamics of Ligand Binding to a Specific Inactive Conformation in Protein Kinases 153
Sanjay B. Hari, Pratistha Ranjitkar, and Dustin J. Maly

13 Purification and Specific Assays for Measuring APE-1 Endonuclease Activity . . 161
Adrian Esqueda, Mohammed Z. Mohammed, Srinivasan Madhusudan, and Nouri Neamati

14 An In Vitro Screening to Identify Drug-Resistant Mutations for Target-Directed Chemotherapeutic Agents 175
Mohammad Azam

15 Utilizing AntagomiR (Antisense microRNA) to Knock Down microRNA in Murine Bone Marrow Cells 185
Chinavenmeni S. Velu and H. Leighton Grimes

16 Synthesis, Conjugation, and Labeling of Multifunctional pRNA Nanoparticles for Specific Delivery of siRNA, Drugs, and Other Therapeutics to Target Cells 197
Peixuan Guo, Yi Shu, Daniel Binzel, and Mathieu Cinier

17 Mouse Models for Tumor Metastasis 221
Shengyu Yang, J. Jillian Zhang, and Xin-Yun Huang

Index *229*

Contributors

MOHAMMAD AZAM • *Divisions of Pathology, Hematology and Cancer Biology, Cancer and Blood Disease Institute, Cincinnati Children's Hospital and Medical Center, Cincinnati, OH, USA; Department of Medicine, University of Cincinnati, Cincinnati, OH, USA*

JACEK BIESIADA • *Biomedical Informatics, Children's Hospital Research Foundation, Cincinnati, OH, USA; Division of Management and Informatics, Technical University of Silesia, Katowice, Poland*

DANIEL BINZEL • *Nanobiomedical Center, University of Cincinnati, Cincinnati, OH, USA*

RICARDO M. BIONDI • *Department of Internal Medicine I, Research Group PhosphoSites, Universitätsklinikum Frankfurt, Frankfurt, Germany*

ANNE BLANGY • *Universités Montpellier2 & 1, Montpellier, France; CNRS, UMR5237, Montpellier, France*

NATHALIE BOUQUIER • *Centre de Recherche de Biochimie Macromoléculaire, CNRS–UMR 5237, Universités Montpellier I et II, Montpellier Cedex, France*

LAN CHEN • *Department of Neurology, Indiana University School of Medicine, Indianapolis, IN, USA*

MATHIEU CINIER • *Nanobiomedical Center, University of Cincinnati, Cincinnati, OH, USA*

CHARLES S. CRAIK • *Department of Pharmaceutical Chemistry, University of California, San Francisco, San Francisco, CA, USA*

ANNE DEBANT • *Centre de Recherche de Biochimie Macromoléculaire, CNRS–UMR 5237, Universités Montpellier I et II, Montpellier Cedex 5, France*

ADRIAN ESQUEDA • *Laboratory of Molecular Oncology, School of Molecular Medical Sciences, University of Nottingham, Nottingham, UK*

PHILIPPE FORT • *Universités Montpellier2 & 1, CRBM, Montpellier, France; CNRS, UMR5237, Montpellier, France*

SYLVIE FROMONT • *Centre de Recherche de Biochimie Macromoléculaire, CNRS–UMR 5237, Universités Montpellier I et II, Montpellier Cedex 5, France*

KALLE GEHRING • *Department of Biochemistry and the Cancer Center, McGill University, Montreal, QC, Canada; Quebec/Eastern Canada High Field NMR Facility, McGill University, Montreal, QC, Canada*

H. LEIGHTON GRIMES • *Immunobiology Division, Cincinnati Children's Hospital Medical Center, Cincinnati, OH, USA*

PEIXUAN GUO • *Nanobiomedical Center, University of Cincinnati, Cincinnati, OH, USA*

SANJAY B. HARI • *Department of Chemistry, University of Washington, Seattle, WA, USA*

VALERIE HINDIE • *Department of Internal Medicine I, Research Group PhosphoSites, Universitätsklinikum Frankfurt, Frankfurt, Germany*

XIN-YUN HUANG • *Department of Physiology, Cornell University Weill Medical College, New York, NY, USA*
DAVID S. LAWRENCE • *Division of Medicinal Chemistry and Natural Products, Department of Chemistry, University of North Carolina, Chapel Hill, NC, USA*
LAURA A. LOPEZ-GARCIA • *Department of Internal Medicine I, Research Group PhosphoSites, Universitätsklinikum Frankfurt, Frankfurt, Germany*
SRINIVASAN MADHUSUDAN • *Laboratory of Molecular Oncology, School of Molecular Medical Sciences, University of Nottingham, Nottingham, UK*
DUSTIN J. MALY • *Department of Chemistry, University of Washington, Seattle, WA, USA*
JAROSLAW MELLER • *Biomedical Informatics, Children's Hospital Research Foundation, Cincinnati, OH, USA; Department of Environmental Health, University of Cincinnati College of Medicine, Cincinnati, OH, USA; Department of Informatics, Nicholas Copernicus University, Torun, Poland*
MOHAMMED Z. MOHAMMED • *Laboratory of Molecular Oncology, School of Molecular Medical Sciences, University of Nottingham, Nottingham, UK*
NOURI NEAMATI • *Department of Pharmacology and Pharmaceutical Sciences, University of Southern California, School of Pharmacy, Los Angeles, CA, USA*
JEFFREY R. PETERSON • *Cancer Biology Program, Fox Chase Cancer Center, Philadelphia, PA, USA*
ALEKSEY POROLLO • *Department of Environmental Health, University of Cincinnati College of Medicine, Cincinnati, OH, USA*
PRATISTHA RANJITKAR • *Department of Chemistry, University of Washington, Seattle, WA, USA*
H. URI SARAGOVI • *Lady Davis Institute-Jewish General Hospital, Montréal, QC, Canada; Departments of Pharmacology and Therapeutics, Oncology and the Cancer Center, McGill University, Montréal, QC, Canada*
SUSANNE SCHMIDT • *Centre de Recherche de Biochimie Macromoléculaire, CNRS–UMR Universités Montpellier I et II, Montpellier Cedex, France*
TINA SHAHIAN • *Graduate Group in Biochemistry and Molecular Biology, University of California, San Francisco, San Francisco, CA, USA*
JUFANG SHAN • *Department of Structural Biology, St. Jude Children's Research Hospital, Memphis, TN, USA*
XUN SHANG • *Division of Experimental Hematology and Cancer Biology, Children's Hospital Medical Center, University of Cincinnati, Cincinnati, OH, USA*
YI SHU • *Nanobiomedical Center, University of Cincinnati, Cincinnati, OH, USA*
TARA SPRULES • *Quebec/Eastern Canada High Field NMR Facility, McGill University, Montreal, QC, Canada*
WENYONG TONG • *Lady Davis Institute-Jewish General Hospital, Montréal, QC, Canada; Departments of Pharmacology and Therapeutics, and the Cancer Center, McGill University, Montreal, QC, Canada*
CHINAVENMENI S. VELU • *Immunobiology Division, Cincinnati Children's Hospital Medical Center, Cincinnati, OH, USA*
JULIEN VIAUD • *Cancer Biology Program, Fox Chase Cancer Center, Philadelphia, PA, USA*

SHENGYU YANG • *Department of Physiology, Cornell University Weill Medical College, New York, USA; Department of Molecular Oncology, Moffitt Cancer Center, Tampa, FL, USA*

J. JILLIAN ZHANG • *Department of Physiology, Cornell University Weill Medical College, New York, NY, USA*

SHENG ZHANG • *Department of Biochemistry and Molecular Biology, Indiana University School of Medicine, Indianapolis, IN, USA*

ZHONG-YIN ZHANG • *Department of Biochemistry and Molecular Biology, Indiana University School of Medicine, Indianapolis, IN, USA*

JIE J. ZHENG • *Department of Structural Biology, St. Jude Children's Research Hospital, Memphis, TN, USA*

YI ZHENG • *Division of Experimental Hematology and Cancer Biology, Cincinnati Children's Hospital Medical Center, University of Cincinnati, Cincinnati, OH, USA*

Chapter 1

On Setting Up and Assessing Docking Simulations for Virtual Screening

Jacek Biesiada, Aleksey Porollo, and Jaroslaw Meller

Abstract

Small molecule docking and virtual screening of candidate compounds have become an integral part of drug discovery pipelines, complementing and streamlining experimental efforts in that regard. In this chapter, we describe specific software packages and protocols that can be used to efficiently set up a computational screening using a library of compounds and a docking program. We also discuss consensus- and clustering-based approaches that can be used to assess the results, and potentially re-rank the hits. While docking programs share many common features, they may require tailored implementation of virtual screening pipelines for specific computing platforms. Here, we primarily focus on solutions for several public domain packages that are widely used in the context of drug development.

Key words: Drug discovery, Virtual screening, Small molecule docking, Flexible docking, AutoDock, Polyview-MM, Visualization, Clustering, Re-ranking

1. Introduction

Virtual screening (VS) can accelerate drug discovery and development by facilitating rapid identification of hits and lead compounds. In a typical application of VS, docking simulations are performed for a large number of candidate compounds in order to rank them according to their predicted binding affinity, and to select top hits for further assessment and validation (1–9). Such simulations require three basic ingredients: (a) a structure of the target receptor; (b) a docking program; and (c) a suitable library of compounds. A computational cluster, or another platform for distributed computing, is typically used to speed up the simulations, often requiring tailored solutions for each program and specific computational platforms. In addition, software packages for molecular modeling, structure visualization, evaluation of docking simulations, and cheminformatic analyses are typically integrated into VS pipelines (1–3).

Yi Zheng (ed.), *Rational Drug Design: Methods and Protocols*, Methods in Molecular Biology, vol. 928, DOI 10.1007/978-1-62703-008-3_1, © Springer Science+Business Media New York 2012

In this chapter, we discuss each of these components with the goal of illustrating practical aspects of setting up and assessing docking simulations for VS. An emphasis is put on publicly available tools that can be used to facilitate subsequent steps in a typical application of VS. A rather informal way of presenting the material is designed to make this chapter accessible to basic scientists without computational training. For the sake of completeness, we start by briefly revisiting some key concepts that underlie the discussion in the subsequent sections of the chapter. The reader is also referred to a number of excellent overviews of the subject (1–3).

Computational docking (and hence VS) requires a high-resolution structure or sufficiently accurate model of a target protein (or another macromolecular receptor). In general, these could be structures resolved by X-ray crystallography or NMR spectroscopy, and available from the Protein Data Bank, or obtained using some modeling approach, e.g., to simulate functionally relevant conformational changes (4). The quality and choice of a receptor structure are likely to have an impact on the results of VS. In particular, docking methods introduce many approximations to simplify the problem, such as the rigid body docking model in which the receptor structure is fixed. Rigid body docking speeds-up computations compared with flexible docking (in which the receptor structure is allowed to move) by precomputing forces experienced by the ligand on a grid. On the other hand, it makes the choice of receptor conformation(s) to be used for docking simulations especially important (even if limited active space consisting of flexible residues is allowed).

Given a receptor structure, docking simulations can be performed using one (or more) of many available docking programs, which has to be selected (and tuned up) carefully when setting up VS pipelines. Conceptually, docking simulations involve two main components, i.e., sampling and scoring, which are implemented in different programs using many alternative approaches. Sampling algorithms are used to find plausible conformations of the receptor–ligand complex, while scoring functions are required to estimate relative binding affinities and rank ligand poses (conformations of the ligand bound to the receptor) (1–9).

The above-mentioned two components are in fact coupled, as the sampling of alternative conformation is embedded into the search for the optimal solution, i.e., the pose with the highest binding affinity (or score). The latter step involves using various optimization techniques, such as Monte Carlo, simulated annealing, or genetic algorithms, and typically requires setting up some depth of search parameter (1). In order to provide the basis for scoring and ranking, atomic force fields and simplified solvation potentials are typically combined into empirical scoring functions (7). It should be emphasized that such effective combinations introduce many approximations to describe both intra- and intermolecular interactions in the system (4) and to estimate the strength of interactions between the ligand and receptor (6). As a result, different scoring functions may introduce distinct biases that have to be taken into account when selecting one of the available docking methods (8).

There are many packages for computational docking (of small molecules) that can be used in the context of VS, including AutoDock (10, 11), DOCK (12–14), Flex (15), Glide (16), GOLD (17), RosettaDock (18), SLIDE (19, 20), and Surflex (21). Stimulated by both methodological advances and fast changes in computing architectures, docking methods have improved considerably over the last decade (22–29). However, benchmarking of docking packages, e.g., using sets of compounds with experimentally measured binding affinities, or assessing enrichment into true positives in VS, suggests that no single method outperforms consistently other methods (24–26). Therefore, different targets may require tailored combinations of several methods, potentially enhanced by rescoring approaches to further limit significant false positive and false negative rates observed in VS (2, 6, 27). Considering a consensus approach utilizing multiple programs can offer another strategy to identify more reliably true binders (24).

In some applications, a relatively small library of carefully selected candidate compounds with specific properties could be selected, e.g., using similarity searches or other cheminformatic approaches. In many cases, though, a wide (essentially unbiased) screening is performed, e.g., limiting the search to drug-like molecules. Thus, a potentially large set of (up to millions of) small molecules that cover most of the pharmacologically interesting subspace of the chemical universe could be considered. As a consequence, computational efficiency of VS is clearly an important issue. Various libraries and data sets of small molecules with desired properties are available for VS, including ZINC (30) and other libraries in public domain. Small molecules included in such databases are typically assessed by using cheminformatic approaches, e.g., to predict their toxicity, pharmacokinetic and other properties that may be relevant for further drug development (7–9).

The overall flow of VS, with docking simulations performed for a number of compounds from some *in silico* library, is illustrated in Fig. 1. Assuming that preprocessing steps, such as the grid preparation, can be performed independently, the overall time required to perform VS is given by the product of the number of compounds and the average time of performing docking for an individual compound. The latter depends on the computing architecture, docking method, and depth of search, and it could take up to hours of CPU time on current multicore processors per compound to achieve sufficient sampling. Thus, as the size of the library increases, it is critical to optimize VS pipelines by taking advantage of distributed computing platforms, and by suing “embarrassing parallelism” of VS (28, 29). In particular, subsets of compounds

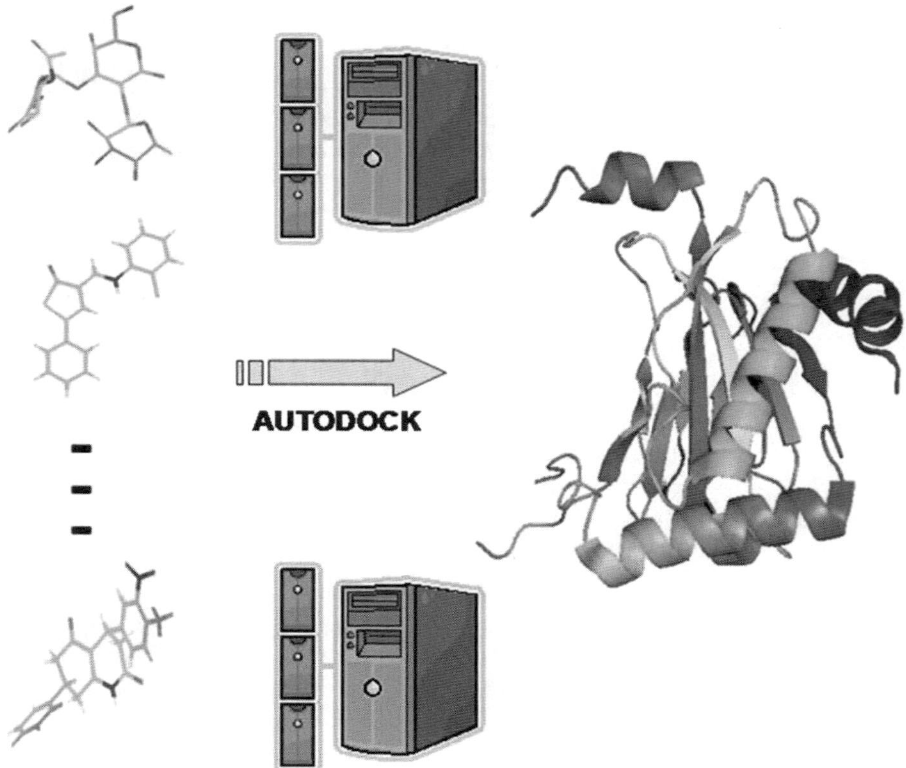

Fig. 1. The overall flow of VS for a library of compounds, with docking simulations performed on a computational cluster and the AutoDock program.

can simply be screened on individual cluster nodes (or some other distributed platform), offering a speedup factor essentially equal to the number of available nodes. Not surprisingly, VS runs are typically being performed on computational clusters, as indicted in Fig. 1. For further details, the reader is also referred to a recent survey of public domain methods packages (25), which was used as a basis for the selection of some topics presented here.

2. Materials

The AutoDock (10, 11) and SLIDE (19, 20) docking programs were used here to illustrate different aspects of docking simulations. Two different libraries of small molecules were used to test VS pipelines discussed later, namely, ZINC (30) and a diverse library of about 300,000 drug-like compounds developed by the former Pharma Division of the PNG company, which was subsequently donated to the Drug Development Center of the University of Cincinnati. The latter library will be referred to as PNG library.

In addition, a number of software programs for visualization and analysis of the results of VS were used, including Polyview-MM, as discussed in Subheading 3.

3. Methods

3.1. Analysis, Modeling, and Selection of the Receptor Structure

Setting up of docking simulations starts with the selection and preparation of a suitable structure of a target receptor (which we assume here is a protein). This step is typically associated with some structure similarity searches, elements of molecular modeling, and various structural analyses. The latter may involve identification of pockets, cavities, interaction interfaces, and specific hot spots within them as targets for small molecules. Molecular modeling steps could be required to build a model of the receptor, e.g., using homology modeling and a resolved template with a sufficiently high degree of homology, or to characterize conformational changes and/or slow coordinated motions in the receptor by using Molecular Dynamics and/or Normal Mode Analysis (4). These steps can be accomplished using many publicly available tools developed by the community.

The primary input file for docking programs usually includes a target protein coordinate file in the PDB or another commonly used format, plus supplementary files that define a site to be targeted by small molecules (or other moieties). Many macromolecular visualization programs, such as Jmol (31), PyMol (32), or VMD (33), can be used to analyze the targeted structure, and such visual inspection and initial analysis can be very helpful in the process of setting up the simulations. For example, if the goal is to design an inhibitor competing with a known ligand, and a high-resolution protein–ligand complex structure is available, as a first step (that ignores induced fit effects) one can simply use an "apo" form obtained by removing the ligand from the complex structure (see Fig. 2). Furthermore, programs such as LIGPLOT (34) can be used to generate a detailed information on protein–ligand interactions, observed distances, and types of interactions according to atoms and molecular groups involved. Analysis of the binding site and contacts between the natural ligand and its receptor may help not only define better the target site but also suggest properties of potential inhibitors.

In general, though, the identification of an optimal binding site to be targeted by small molecules can be challenging. For example, when protein–protein interaction interfaces are targeted to disrupt protein interactions, hot spots that contribute significantly to the binding energy, surface cavities, or sites that contribute to functionally relevant hinge motions or other conformational

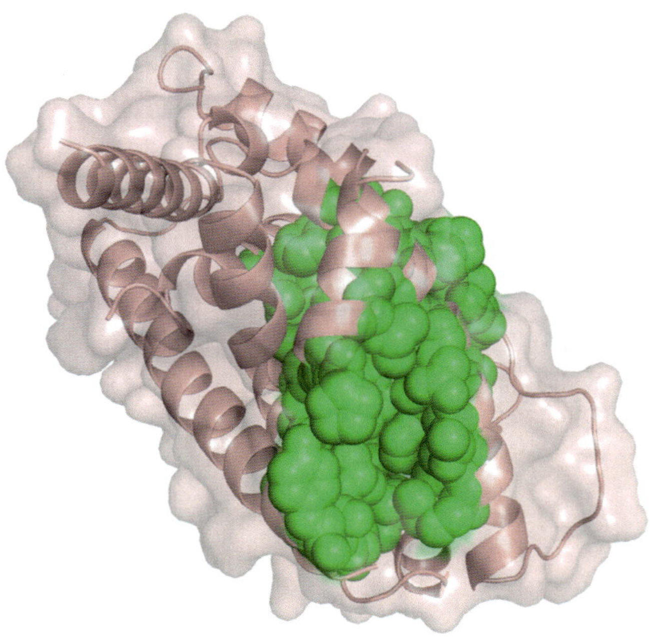

Fig. 2. Ligand-binding pocket in the ERα ligand-binding domain (PDB ID: 1ERR), as identified using CASTp, and shown using *green spheres*. Automated analysis and annotation of binding pockets and cavities retrieved from CASTp are available through the Polyview-3D server.

changes, such as those associated with induced fit upon ligand binding, may need to be identified and characterized. In this context, it should be noted that structural ensemble-based and flexible docking methods can in principle account for induced fit and other conformational changes. However, increased computational cost and limited accuracy may in practice require that considering a number of carefully selected conformations within the rigid body approximation may be more feasible.

Sequence- and structure-based mapping of putative hot spots and functional sites to be targeted can be coupled with further assessment of their druggability (35, 36). On the other hand, slow coordinated motions of the receptor can be modeled by using Elastic Network Model (ENM) or Molecular Dynamics (MD)-based normal mode analysis (37, 38). Example of such analysis is illustrated in Fig. 3, which was used for initial analysis of a hinge motion involving two domains of a viral capsid protein, with the goal of identifying intermediate conformations that could potentially be targeted to destabilize the capsid.

While many tools exist to facilitate such analyses, many of these steps can be accomplished using the Polyview-3D server (http://polyview.cchmc.org/polyview3d.html) (39) that combines high-quality structure visualization and animation with a range of structure annotation tools. The latter include ConSurf (40) for mapping

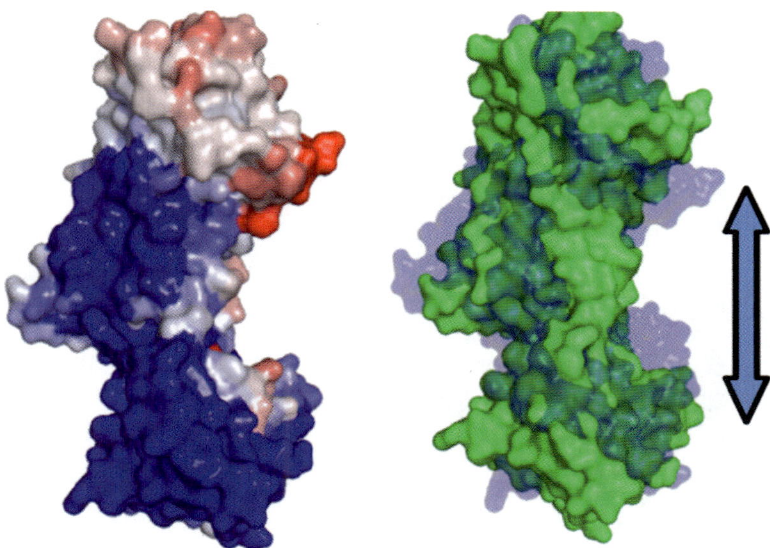

Fig. 3. Analysis of local fluctuations (*left panel*) and slow coordinated motions (*right panel*) in the Norwalk virus capsid protein (PDB ID: 1IHM). The extent of local fluctuations is shown by mapping temperature (or B-) factors from the crystal structure (with low B-factors shown in *blue* vs. high B-factors that correspond to flexible sites shown in *red*). On the other hand, ENM analysis of slow coordinated hinge-type motion of the two domains of the capsid protein (indicated by the *arrow* in the *right panel*) was obtained using the Nomad-ref server (*blue* transparent structure in the *right panel* represents a structure distorted along one of the normal modes). Pictures were generated using the Polyview-3D server.

evolutionary conserved sites in proteins, CASTp (41) for analysis of pockets and cavities (an example is included in Fig. 2), and SPPIDER (42) for mapping and prediction of protein interaction interfaces. Each of these analyses is made available through a simple interface, automating tedious intermediate steps and allowing nonexperts to easily generate the results (see also Polyview-3D tutorial available from the server).

3.2. Docking Simulations with AutoDock

In what follows, we primarily use AutoDock (11) as an example of a docking program in order to illustrate some further specific issues and practical consideration arising in the context of VS. The popularity of AutoDock, which is one of the most widely used public domain packages for docking, stems from its well-benchmarked state-of-the-art performance, availability of versatile implementations that evolve to take advantage of emerging new computing platforms (as discussed in more detail later), and the fact that it is available for free for academic use (http://autodock.scripps.edu/).

AutoDock includes several established force fields, recently (as of version 4.2) extended by the Amber molecular force field. Both efficient (grid-based) implementation of rigid body docking and flexible (receptor) docking protocols, with an active space consisting of residues that are allowed to undergo conformational changes, are also available. The search for ligand poses/complex conformations with high binding affinity is performed using an efficient implementation of Lamarckian genetic algorithm, with sampling and "mixing" of suboptimal solutions to solve the underlying global optimization problem (11).

A typical docking process with AutoDock consists of the following main steps (shared with many other docking packages): (a) preparing the ligand (or a library of ligands in the case of VS) represented by files describing its covalent structure and force field parameters, such as partial charges; (b) preparing the protein receptor and setting up the simulation system, which involves preparing a grid box covering the binding site of interest and generating the corresponding map files; (c) performing the actual docking simulations with some depth search as defined by relevant parameters; and (d) post-processing and analysis of the results. These steps are schematically represented in a flowchart of the docking pipeline in Fig. 4.

When performing docking simulations for a single ligand (as often done in re-docking experiments to test the parameter setup, convergence of sampling, or to benchmark different methods), all the steps listed above can be achieved using AutoDockTools (ADT) package that was specifically developed to facilitate setting up simulations with AutoDock (11). In particular, ADT allows one to define a grid box that contains the targeted binding site, and to generate map files that store pre-computed forces to be used in the course of simulations. Another tool that provides an interface for

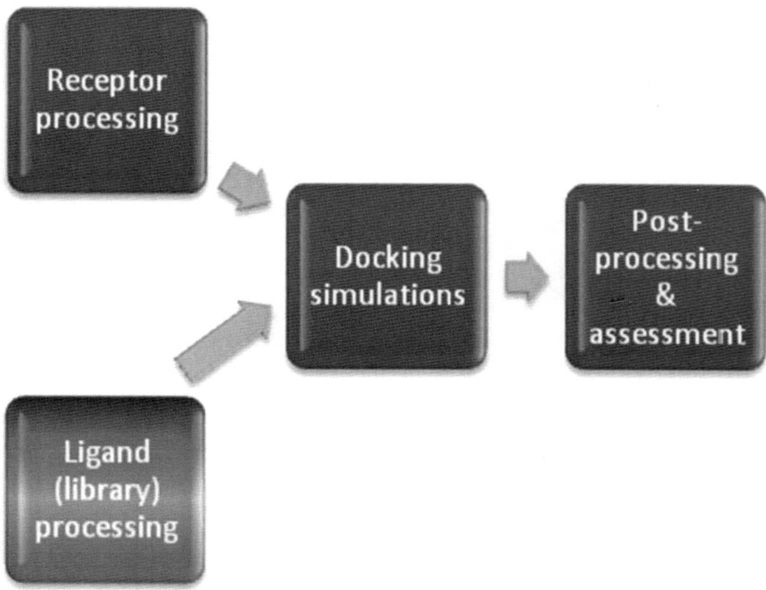

Fig. 4. Schematic flowchart of a typical virtual screening pipeline, with different stages of VS represented by *rectangular boxes*.

setting up simulations with AutoDock is vsLab (43). ADT and vsLab can also be used to analyze the results of docking simulations, as discussed later.

When considering a VS run for a library of compounds, an interactive interface can still be used to prepare the receptor structure, which is typically the same for all compounds screened. On the other hand, preprocessing and setting up simulations for all (or some) compounds in a library require scripts or other automated solutions. There are several packages that can be used for that purpose that have been developed to work with AutoDock (often also applicable to other programs with minimal changes), including PyRx (44) and Raccoon (45). The latter provides a user-friendly interface (Python GUI) and enables automated preparation of *in silico* libraries, including conversion to AutoDock pdbqt format, generation of scripts for cluster submission using commonly used PBS queuing system, and post-processing of results. When using Raccoon, the map (grid) files for AutoDock are created only once for all possible atoms in the library, as part of preprocessing step, and thus maybe reused for all ligands in the library for docking process. This is highlighted by using a different color for the library preparation box in Fig. 4.

In order to streamline docking simulations and VS, we have developed a set of scripts with same main objective and functionality as Raccoon, which provides us with additional flexibility and tailored solutions when using multiple docking programs (e.g., AutoDock, Vina, and SLIDE) and hierarchical, multistep VS process with increasing search depth, as illustrated in Table 1. In particular, we introduced an index file that can be used to define any

Table 1
A multistage protocol for VS of large libraries, using AutoDock with increasing depth search to improve the quality of sampling, and decreasing grid spacing to improve the accuracy of the force computation

	ga_num_evals	ga_pop_size	ga_run	Grid spacing [Ang]	N_compounds
Phase I	250,000	75	20	0.600	300,000
Phase II	1,000,000	75	50	0.375	30,000
Phase III	2,000,000	150	50	0.375	3,000

The effective depth of search is defined by the maximum number of energy evaluations (ga_num_evals), the population size used in genetic algorithm (ga_pop_size), and the number of repeated docking runs (ga_run). These values of these parameters can be contrasted with 50 Lamarckian GA runs, 10 million energy evaluations, and population size of 150 that were used to obtain the results on the ER benchmark discussed in the text. Using such (extensive) sampling, the average running time per compound was 2,987 CPU seconds on the Xeon(R) X5570 2.93 GHz quad-core server. The number of compounds screened in each phase (ranked according to predicted binding affinities in the previous iteration) is given in the last column

subset of the library for VS (to be distributed among computing nodes), greatly simplifying the maintenance of all libraries, which can be all stored in a proper format (pdbqt) in one location, each library in a single file irrespective of the setup of the docking simulations on cluster. In addition, different cluster submission/scheduling systems, including PBS and LSF, can be used.

Preparing the receptor protein for docking simulations consists of several simple steps, including cleaning the coordinate file and removing alternative conformers, removing water molecules, adding hydrogens, and adding partial charges using a script included in ADT (prepare_receptor4.py), or by using interactively the ADT GUI interface. As indicated before, the next important step is the preparation of a grid box, which again can be done using ADT and assuming that a binding site to be targeted has been identified, e.g., in terms of residues that define the location of the box in terms of their geometric center.

The subsequent steps involve preparation of configuration files for cluster submission scripts, where we specify for example the number of nodes to be used in calculation, grid parameters and location of map files, information about library location and index files that define subsets to be distributed on computing nodes, output file location, etc. At this point the calculations can be submitted to the cluster, using the available scheduling system. Our VS pipeline has been extended to add several scripts that facilitate breaking and restarting the calculations, and control the progress of the screening. These capabilities play an important role in monitoring the simulations, as they may be distributed over a substantial number of nodes.

3.3. Analysis and Assessment of the Results

Despite constant progress, the accuracy of current docking methods is limited, resulting in significant rates of both false positives and false negatives in VS. To partially address limitations with regard to scoring functions and conformational sampling, docking packages typically perform multiple docking simulations per ligand and use the analysis of predicted complexes, e.g., in terms of consistency of ligand poses across multiple runs, to assess and potentially re-rank hits for virtual screening. Putative binders can also be assessed in terms of the distribution of the predicted binding affinities observed in repeated docking simulations, which can be characterized in terms of maximum, average, and median inhibition constants or binding energy.

In Fig. 5, we illustrate how median affinity scores can be combined with measures of the consistency (or lack thereof) of docking poses, which is quantified using normalized entropy. A value of 0 corresponds to all poses forming just one cluster, i.e., being essentially identical (which also implies that in all runs the ligand binds to the same binding site), and value of 1 corresponding to each pose forming its own cluster, i.e., each run resulting in binding to

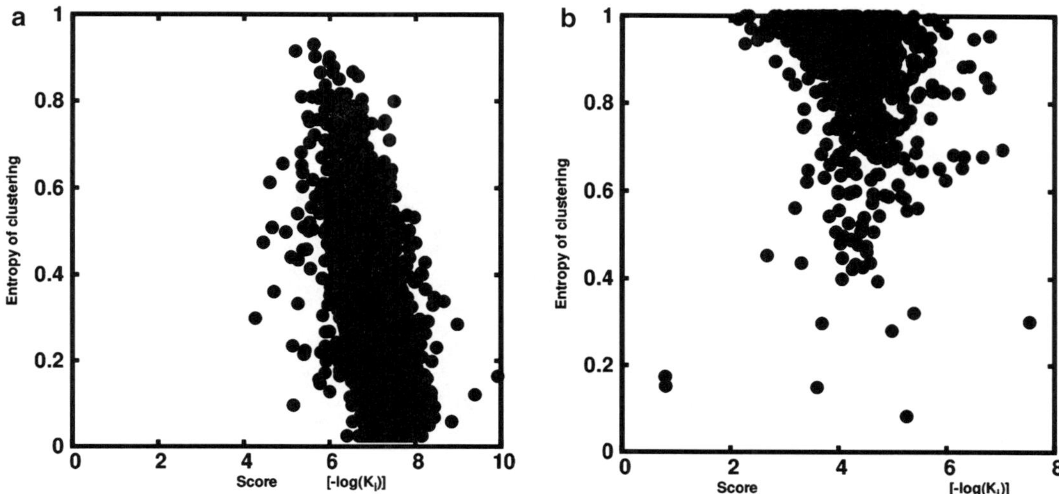

Fig. 5. Assessment of the clustering of docking poses, as measured by the entropy of the distribution of poses over the distinct clusters of poses (obtained in 50 different runs of docking simulations). Results for about 2,500 top hits ranked according to the median inhibition constant for two different sites on the Ras-binding Sos1 protein (PDB code 1XD2) are shown in (**a**) and (**b**), respectively. While hits of potential interest with median inhibition constants K_i in the micromolar (or better) range are observed in both cases, only in the first case individual docking simulations converge to the same binding site for a significant fraction of ligands, as indicated by low values of entropy for those ligands.

a different site, or the same site, but with a significantly different conformation of the ligand.

Obviously, the bigger the initial simulation box (grid), the more likely it is that alternative binding sites with good binding affinities will be found. Therefore, performing additional simulations to test stability of the solutions with bigger boxes, and alternative (or perturbed, e.g., along normal modes) structures, can be used as part of the overall assessment of the results. Analysis of this type can help in reducing false positive rates by eliminating hits that are predicted to have high binding affinity, without much consistency in terms of their binding mode (which at best could indicate nonspecific binders that should be eliminated anyway).

From the point of view of the user, the assessment of the results typically starts with a visual inspection of the predicted binding modes for top hits. For larger sets of ligands, one needs an automated pipeline that involves tools for structural analysis, geometric clustering of predicted poses, rescoring, and cheminformatic approaches. Below, we discuss several tools that can be either used interactively or incorporated into automated (script-based) post-processing pipelines.

The output from ligand docking programs usually includes a coordinate file (in the PDB format or other standard formats) and supplementary data, such as estimated binding affinity, clustering of alternative poses, etc. To view coordinate files, commonly used macromolecular visualization programs that were discussed before

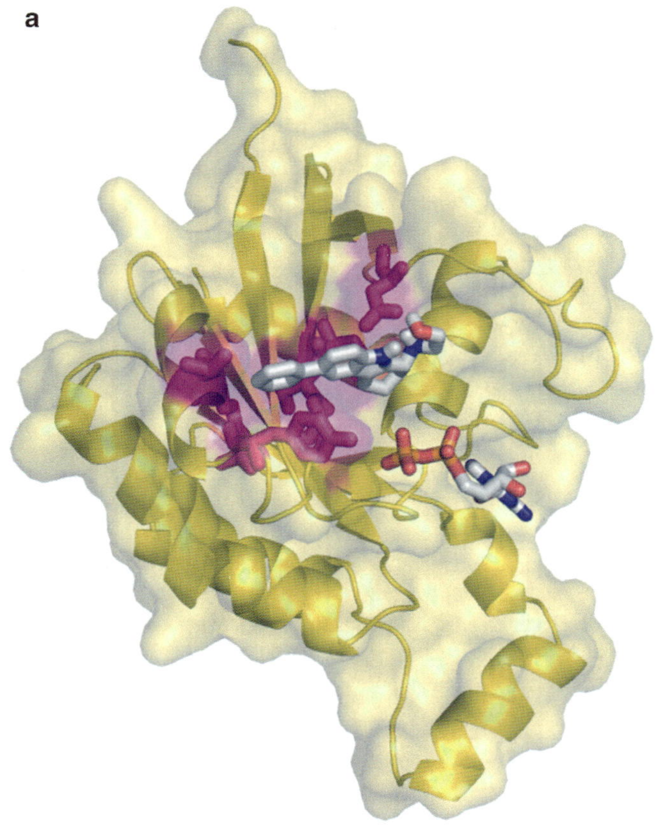

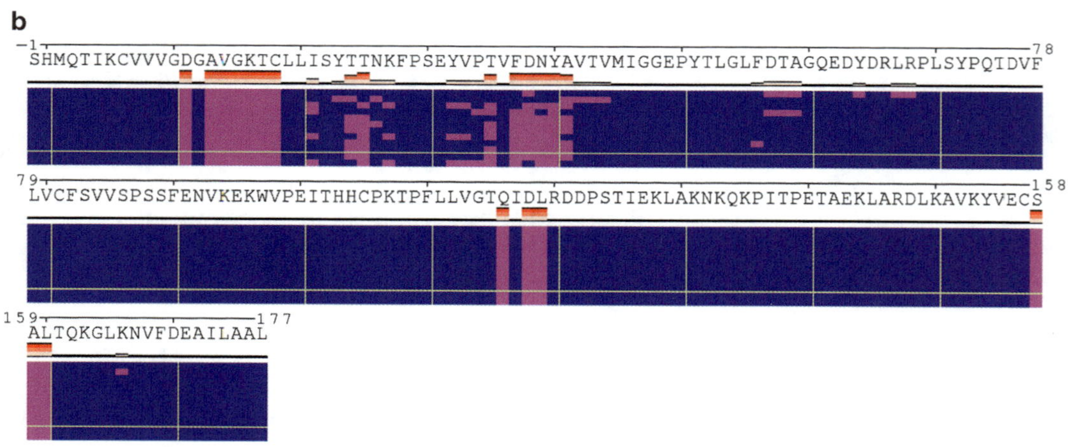

Fig. 6. Analysis of docking poses using Polyview-MM. Residues in contact with a ligand, docked using AutoDock to a small GDP-bound GTPase (PDB ID: 2WMN), are shown in *magenta* in (**a**). On the other hand, (**b**) shows a composite picture of all docking poses, as clustered by AutoDock, with each row in the figure representing one cluster. *Magenta* is again used to show residues in contact with the ligands (including GDP in this case). As can be seen from the figure, several residues are always in contact with a ligand which is always present at one site (GDP), while the binding site for the docked molecule changes from one pose to another, as indicated by the variable pattern of *magenta* sites.

can be used. Some tools discussed here are coupled with a specific docking program to facilitate setting up simulations and subsequent analysis of the results for that program. On the other hand, tools (or options) that simply require one of the standard file formats, such as the PDB format, can easily be incorporated into pipelines for the assessment of VS.

For docking simulations performed using AutoDock, ADT mentioned before in the context of the analysis of the receptor structure and setting up the simulations also provides a number of options for the analysis of the results starting from standard (*.dlg) output files. In particular, ADT offers the possibility to (interactively) view clustered poses of a ligand, characterize individual clusters, and compute RMSD difference with respect to the input (reference) structure. ADT can also represent ligands with the iso-contour rendering to further facilitate the evaluation of individual poses in terms of interatomic interactions between the receptor and the ligand. vsLab can also be used for analyzing docking simulations with AutoDock, with results summarized using both easy-to-navigate tabular views and visualization of docking poses by VMD. In addition to similar visualization capabilities, DockingServer (46) can be used to assess the forces driving specific protein–ligand interactions, whereas ViewDock (47) enables interactive analysis of multiple ligands docked to the receptor.

The last tool that we would like to mention in this context is the POLYVIEW-MM server (48) that we specifically designed to facilitate the analysis of the receptor structure, protein–ligand interactions, and docking results. Both static pictures and animations of ligand poses can be generated using the Jmol applet, or the high-quality PyMol rendering. The server also provides automated analysis of AutoDock clustering for the assessment and potential rescoring of docking poses, as illustrated in Fig. 6. Polyview-MM can also be used to map and generate animated movies in which residues in contact with the ligand are highlighted for each pose, and for further structural and functional annotations, available also through Polyview-3D.

4. Notes

Starting form a library of compounds, the goal of VS is to find a set of high-affinity hits of manageable size that can subsequently be validated, using empirical binding and functional/activity assays. In successful applications, the set of top hits is expected to be significantly enriched in true positives, compared to random subsets of compounds. In fact, enrichment analysis for sets of compounds with experimentally known binding affinities that contain both active (binders) and non-active (no or weak binders) offers the possibility of evaluating docking programs in terms of their performance in VS (2, 6, 24–27).

In this regard, we would like to comment that we found significant enrichment in true positives in our recent analysis (25) of AutoDock and other docking programs on a diverse set of potential estrogens (known as NCTER ER (49, 50)), i.e., compounds that either bind to estrogen receptor (ER) with known binding affinity (active compounds) or are classified as inactive. We also observed encouraging correlations between predicted and experimental binding affinities for active compounds (correlation coefficients at the level of about 0.6) (25). However, high false positive rates were also observed. Moreover, these rates and correlations were significantly affected by the choice of the docking method and the receptor structure, underscoring further the importance of the induced fit, as well as inherent limitations of scoring functions and sampling approaches used by different docking programs.

New implementations of docking programs evolve together with advances in computing. For example, an efficient MPI-based implementation of AutoDock has been developed to enable performing docking simulations for an individual compound in parallel as well (and thus going beyond "embarrassing parallelism") (29). Another example is the development of a CUDA-enabled version of AutoDock (51), which utilizes graphics processing units (GPUs). Recently, multicore servers with hyperthreading have become increasingly popular as a scientific computing platform. In the recently developed AutoDock Vina package (28), which maintains input and output file compatibility with AutoDock, the speed of docking simulations is increased by using a search strategy specifically optimized for hyperthreading.

These new approaches, coupled with constant improvements in hardware, are likely to bring about further significant speedups. At the same time, improvements in the overall accuracy are also expected by enabling more extensive sampling and tuning-up of the parameters. More comprehensive assessment of the impact of alternative receptor structures and docking protocols, in particular those that explicitly account for the receptor flexibility, is also expected to contribute to successful applications of VS in drug design.

Acknowledgments

This work was supported in part by NIH grants A1055649, UL1RR026314, and P01HD013021. Computational resources were made available by Cincinnati Childrens Hospital Research Foundation and University of Cincinnati College of Medicine.

References

1. Huang SY, Zou X (2010) Advances and challenges in protein-ligand docking. Int J Mol Sci 11:3016–3034
2. Ripphausen P, Nisius B, Bajorath J (2011) State-of-the-art in ligand-based virtual screening. Drug Discov Today 16(9–10):372–376
3. Morris GM, Lim-Wilby M (2008) Molecular docking. Methods Mol Biol 443:365–382
4. Petrenko R, Meller J (2009) Molecular dynamics. In: Encyclopedia of life sciences. Wiley
5. Kellenberger E, Rodrigo J, Muller P, Rognan D (2004) Comparative evaluation of eight docking tools for docking and virtual screening accuracy. Proteins 57:225–242
6. Rajamani R, Good AC (2007) Ranking poses in structure-based lead discovery and optimization: current trends in scoring function development. Curr Opin Drug Discov Devel 10:308–315
7. Duch W, Swaminathan K, Meller J (2007) Artificial intelligence approaches for rational drug design and discovery. Curr Pharm Des 13:1497–1508
8. Warren GL, Andrews CW, Capelli AM, Clarke B et al (2006) A critical assessment of docking programs and scoring functions. J Med Chem 49:5912–5931
9. Kolb P, Ferreira RS, Irwin JJ, Shoichet BK (2009) Docking and chemoinformatic screens for new ligands and targets. Curr Opin Biotechnol 20:429–436
10. Goodsell DS, Morris GM, Olson AJ (1996) Automated docking of flexible ligands: applications of AutoDock. J Mol Recognit 9:1–5
11. Morris GM, Huey R, Lindstrom W, Sanner MF et al (2009) AutoDock4 and AutoDockTools4: Automated docking with selective receptor flexibility. J Comput Chem 30:2785–2791
12. Lang PT, Brozell SR, Mukherjee S, Pettersen EF et al (2009) DOCK 6: combining techniques to model RNA-small molecule complexes. RNA 15:1219–1230
13. Shoichet BK, Bodian DL, Kuntz ID (1992) Molecular Docking Using Shape Descriptors. J Comput Chem 13:380–397
14. Meng EC, Shoichet BK, Kuntz ID (1992) Automated Docking with Grid-Based Energy Evaluation. J Comput Chem 13:505–524
15. Claussen H, Buning C, Rarey M, Lengauer T (2001) FlexE: efficient molecular docking considering protein structure variations. J Mol Biol 308:377–395
16. Friesner RA, Banks JL, Murphy RB, Halgren TA et al (2004) Glide: a new approach for rapid, accurate docking and scoring 1. Method and assessment of docking accuracy. J Med Chem 47:1739–1749
17. Verdonk ML, Cole JC, Hartshorn MJ, Murray CW et al (2003) Improved protein-ligand docking using GOLD. Proteins 52:609–623
18. Davis IW, Baker D (2009) RosettaLigand docking with full ligand and receptor flexibility. J Mol Biol 385:381–392
19. Zavodszky MI, Sanschagrin PC, Korde RS, Kuhn LA (2002) Distilling the essential features of a protein surface for improving protein-ligand docking, scoring, and virtual screening. J Comput Aided Mol Des 16: 883–902
20. Zavodszky MI, Rohatgi A, Van Voorst JR, Yan H et al (2009) Scoring ligand similarity in structure-based virtual screening. J Mol Recognit 22:280–292
21. Jain AN (2003) Surflex: fully automatic flexible molecular docking using a molecular similarity-based search engine. J Med Chem 46: 499–511
22. Huang SY, Zou X (2007) Ensemble docking of multiple protein structures: considering protein structural variations in molecular docking. Proteins 66:399–421
23. Morris GM, Huey R, Olson AJ (2008) Using AutoDock for ligand-receptor docking. Curr Protoc Bioinformatics 24:8.14.1–8.14.40
24. Yang JM, Chen YF, Shen TW, Kristal BS et al (2005) Consensus scoring criteria for improving enrichment in virtual screening. J Chem Inf Model 45:1134–1146
25. Biesiada J, Porollo A, Velayutham P, Kouril M, Meller J (2011) Survey of public domain software for docking simulations and virtual screening. Hum Genomics 5(5):497–505
26. Kirchmair J, Markt P, Distinto S, Wolber G et al (2008) Evaluation of the performance of 3D virtual screening protocols: RMSD comparisons, enrichment assessments, and decoy selection–what can we learn from earlier mistakes?, J Comput Aided Mol Des 22:213–228
27. Kim R, Skolnick J (2008) Assessment of programs for ligand binding affinity prediction. J Comput Chem 29:1316–1331
28. Trott O, Olson AJ (2010) AutoDock Vina: improving the speed and accuracy of docking with a new scoring function, efficient optimization, and multithreading. J Comput Chem 31:455–461
29. Khodade P, Prabhu R, Chandra N, Raha S et al (2007) Parallel implementation of AutoDock. J Appl Crystallogr 40:598–599
30. Irwin JJ, Shoichet BK (2005) ZINC–a free database of commercially available compounds

for virtual screening. J Chem Inf Model 45: 177–182

31. Jmol: an open-source Java viewer for chemical structures in 3D. http://www.jmol.org/
32. DeLano WL http://www.pymol.org/
33. Humphrey W, Dalke A, Schulten K (1996) VMD: visual molecular dynamics. J Mol Graph 14:33–38, 27–38
34. Wallace AC, Laskowski RA, Thornton JM (1995) LIGPLOT: a program to generate schematic diagrams of protein-ligand interactions. Protein Eng 8:127–134
35. Seco J, Luque FJ, Barril X (2009) Binding site detection and druggability index from first principles. J Med Chem 52(8):2363–2371
36. Schmidtke P, Barril X (2010) Understanding and predicting druggability. A high-throughput method for detection of drug binding sites. J Med Chem 53(15):5858–5867
37. Cui Q, Bahar I (2006) Normal mode analysis: theory and applications to biological and chemical systems. Chapman & Hall, Boca Raton
38. Chennubhotla C, Rader AJ, Yang LW, Bahar I (2005) Elastic network models for understanding biomolecular machinery: from enzymes to supramolecular assemblies. Phys Biol 2(4): S173–S180
39. Porollo A, Meller J (2007) Versatile annotation and publication quality visualization of protein complexes using POLYVIEW-3D. BMC Bioinformatics 8(316)
40. Landau M, Mayrose I, Rosenberg Y, Glaser F, Martz E, Pupko T, Ben-Tal N (2005) ConSurf 2005: the projection of evolutionary conservation scores of residues on protein structures. Nucleic Acids Res 33:W299–W302
41. Dundas J et al (2006) CASTp: computed atas of surface topography of proteins with structural and topographical mapping of functionally annotated residues. Nucl Acids Res 34:W116–W118
42. Porollo A, Meller J (2007) Prediction-based fingerprints of protein-protein interactions. Proteins 66:630–645
43. Cerqueira NMFSA, Ribeiro J, Fernandes PA, Ramos MJ (2011) vsLab—An implementation for virtual high-throughput screening using AutoDock and VMD. Int J Quantum Chem 111:1208–1212
44. Wolf LK (2009) New software and Websites for the Chemical Enterprise. Chem Eng News 87:31
45. Forli S Raccoon|AutoDock VS: an automated tool for preparing AutoDock virtual screenings. http://autodock.scripps.edu/resources/raccoon
46. DockingServer. http://www.dockingserver.com/web
47. Pettersen EF, Goddard TD, Huang CC, Couch GS et al (2004) UCSF Chimera–a visualization system for exploratory research and analysis. J Comput Chem 25:1605–1612
48. Porollo A, Meller J (2010) POLYVIEW-MM: web-based platform for animation and analysis of molecular simulations. Nucleic Acids Res 38(Suppl):W662–W666
49. Fang H, Tong W, Shi LM, Blair R et al (2001) Structure-activity relationships for a large diverse set of natural, synthetic, and environmental estrogens. Chem Res Toxicol 14:280–294
50. Barrett I, Meegan MJ, Hughes RB, Carr M et al (2008) Synthesis, biological evaluation, structural-activity relationship, and docking study for a series of benzoxepin-derived estrogen receptor modulators. Bioorg Med Chem 16:9554–9573
51. AutoDock Software in Parallel with GPUs. http://gpuautodock.sourceforge.net/
52. Lindahl E, Azuara C, Koehl P, Delarue M (2006) NOMAD-Ref: visualization, deformation and refinement of macromolecular structures based on all-atom normal mode analysis. Nucleic Acids Res 36:W52–W56

Chapter 2

Virtual Ligand Screening Combined with NMR to Identify Dvl PDZ Domain Inhibitors Targeting the Wnt Signaling

Jufang Shan and Jie J. Zheng

Abstract

Virtual ligand screening is a powerful technique to identify potential hits of targets and to increase hit rates. Here, we describe how we used this technique combined with NMR ^{15}N HSQC experiments to obtain small molecules that bind to the PDZ domain of Dvl targeting the Wnt signaling pathway.

Key words: Virtual ligand screening, Database search, Docking, NMR, ^{15}N HSQC, SAR

1. Introduction

Wnt signaling pathways play key developmental and growth regulatory roles that significantly affect many biological processes; abnormal Wnt activity has been implicated in cancer and other human diseases. Dishevelled (Dvl) relays the signal from the membrane-bound Wnt receptors of the Frizzled (Fz) family to downstream substrates via its PDZ domain (1). The special role of the Dvl PDZ domain in the Wnt signaling pathway (1) makes it a potential pharmaceutical target (2). Small organic inhibitors of the PDZ domain in Dvl (3, 4) might be useful in dissecting molecular mechanisms and formulating pharmaceutical agents that target cancers or other diseases in which Wnt signaling is involved.

1.1. Structure-Based Virtual Ligand Screening for Inhibitor of the Dvl PDZ Domain

Like many other PDZ domains (5), the structure of the Dvl PDZ domain is known (6). This has permitted us to use structure-based virtual ligand screening (VLS) to computationally access potential ligands. By using a UNITY search for compounds with the potential to bind to the PDZ domain, FlexX docking of candidates into the binding site, Cscore ranking of binding modes, and chemical shift perturbation NMR experiments, we identified a small organic

Yi Zheng (ed.), *Rational Drug Design: Methods and Protocols*, Methods in Molecular Biology, vol. 928,
DOI 10.1007/978-1-62703-008-3_2, © Springer Science+Business Media New York 2012

molecule (NSC668036) from the National Cancer Institute (NCI) small-molecule library that can bind to the mDvl1 PDZ domain. Further NMR experiments confirmed that the compound binds to the peptide-binding site on the surface of the PDZ domain. In addition, we carried out molecular dynamics (MD) simulations of the interaction between this compound and the PDZ domain as well as that between the C-terminal region of a known PDZ domain inhibitor (Dapper) and the PDZ domain, and we compared the binding free energies of these interactions calculated via the molecular mechanics Poisson–Boltzmann surface area (MM-PBSA) method (7–9).

1.2. Optimizing Dvl PDZ Domain Inhibitor by Virtually Exploring Chemical Space

With the advancement of technology in the field of drug discovery, hits of a potential therapeutic reagent can be identified in a comparatively straightforward fashion by using high-throughput screening (10, 11). However the follow-up hit-to-lead process and lead optimization still remain as challenging problems in the drug discovery process (12, 13). One of the most frequently taken approaches in the hit-to-lead process is hit evolution (13). During hit evolution, analogues of the most promising hits are synthesized for the development of structure–activity relationship (SAR) data. The SAR is then used to guide the synthesis and optimization of lead compounds to improve their potencies and physicochemical properties, and to reduce off-target activities. The synthesis processes are usually long and labor intensive. Virtual screening of databases consisting of physically available compounds may help us to take advantage of the chemistry that has already been done and speed up projects, especially with the ever-growing list of existing compounds. Indeed, the Zinc database has 13 million compounds (14) and the iResearch™ Library (ChemNavigator, San Diego, CA) has more than 50 million unique chemicals. Although the databases of available compounds are still under-sampled (15), the chemical space represented by those millions of compounds should never be neglected. We believed that the large chemical space of available compounds offers us with an opportunity to explore SAR of known hits; and as a proof of principle test, we searched the ChemDiv database for the Dvl PDZ domain inhibitors based on an inhibitor identified above (16). In our studies, we first developed a pharmacophore model based on NSC668036; based on the model, we then screened the ChemDiv database by using an algorithm that combines similarity search and docking procedures; finally, we selected potent inhibitors based on docking analysis and examined them by using NMR spectroscopy. NMR experiments showed that all the 15 compounds we chose bound to the PDZ domain tighter than NSC668036.

2. Materials

2.1. Virtual Screening

1. The structure of the Dvl PDZ domain was obtained from the protein data bank (PDB entry 1L6O (6)) (Note 1).
2. The three-dimensional (3D) NCI database of small-molecule is available from NCI at no cost, and it includes the coordinates of more than 250,000 drug-like chemical compounds. The Tripos format of NCI database comes with SYBYL® (St Louis, MO) and is ready for UNITY search. For ChemDiv database, UNITY in SYBYL® was used to convert the SLN files obtained from ChemDiv Inc. (San Diego, CA).

2.2. NMR Experiments

1. The ^{15}N-labeled mouse Dvl1 PDZ domain (residues 247–341 of mouse Dvl1) was prepared as described previously (1, 17) by the protein production facility at St. Jude Children's Research Hospital. For NMR, the 0.3 mM ^{15}N-labeled Dvl PDZ domain was in 100 mM potassium phosphate buffer (pH 7.5), 10% D_2O, and 0.5 mM EDTA.
2. Compounds from the NCI library were obtained from Drug Synthesis and Chemistry Branch, Developmental Therapeutics Program, Division of Cancer Treatment and Diagnosis, NCI (http://129.43.27.140/ncidb2/). The 15 ChemDiv compounds were purchased from ChemDiv Inc. (San Diego, CA). Compounds were dissolved in the 100 mM potassium phosphate buffer (pH 7.5) with 10% D_2O, and 0.5 mM EDTA. DMSO was added for compounds not soluble in the above buffer (see Note 3).

3. Methods

3.1. Structure-Based Ligand Screening of Small Compounds Binding to the PDZ Domain

1. The UNITY module of the SYBYL software package (Tripos, Inc.) was used to screen the NCI small-molecule three-dimensional database for chemical compounds that could fit into the peptide-binding groove of the Dvl PDZ domain. Based on the complex structure of PDZ domain and Dapper peptide (6), a search query was designed; it consists of two hydrogen bond donors (backbone amide nitrogens of Gly266 and Ile269) and two hydrogen bond acceptors (carbonyl oxygens of Ile267 and Ile269) on the PDZ domain, with 0.3 Å tolerances for spatial constraints (Fig. 1).
2. The Flex search module of UNITY was used to explore the 3D small-molecule NCI database to identify compounds that met the requirements of the query. The Flex search option of

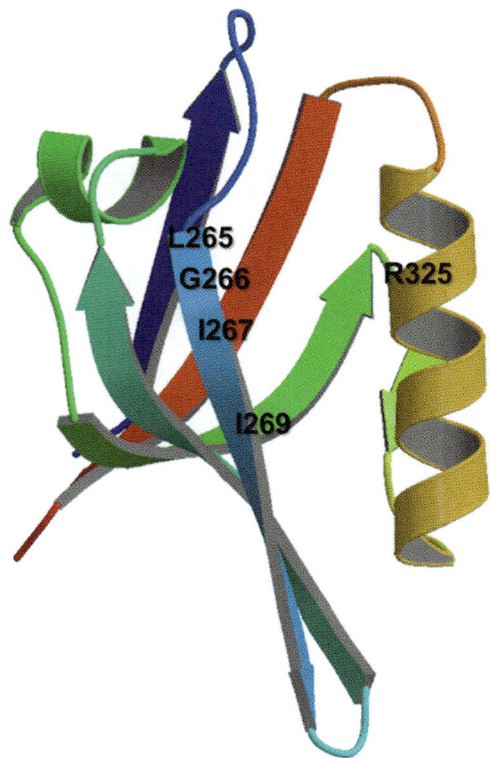

Fig. 1. The structure of the Dvl PDZ domain. The ligand site is located at α-helix B (*yellow*) and β-sheet B (*light blue*) regions.

UNITY considers the flexibility of compounds, and it uses the Directed Tweak algorithm to conduct a rapid and conformationally flexible 3D search (18). The search yielded 108 organic compounds as the initial hits.

3. These 108 hits then were "docked" into the binding site of the PDZ domain by using the FlexX program of SYBYL. FlexX is energy minimization modeling software that varies the conformation of the ligand to fit it into the protein-binding site (19). The receptor's binding site was defined by residues Gly266, Ile269, and Arg325 with a selection radius of 5.9 Å, and a core subpocket was defined by Gly266 with a selection radius of 5.9 Å (Fig. 1). The docking condition was validated by docking the dapper peptide to the PDZ domain as in Note 2.

4. The results of the docking procedure were evaluated manually and those compounds that were not docked into the binding pocket of the PDZ domain were removed. The Cscore program of SYBYL was then used to rank the remaining compounds on the basis of their predicted ability to bind to the binding pocket. Cscore generates a relative, consensus score, based on the individual scoring functions of the protein–ligand complex (20). One of the scoring functions in Cscore, the

F_score, is particularly useful. F_score considers polar and nonpolar interactions in calculating the binding free energy of ligand and protein. Based on this notion, nine compounds whose F_scores were better than that of the control Dapper–PDZ interaction were then further characterized and subsequently obtained from the Developmental Therapeutics Program of the NCI.

3.2. Testing VLS Hits with NMR Spectroscopy

The abilities of the nine compounds obtained from DTP to bind to the PDZ domain were tested by NMR spectroscopy, mainly the chemical shift perturbation experiment (21). NMR ^{15}N HSQC experiments were performed by using a Varian Inova 600 MHz NMR spectrometer at 25°C.

1. HSQC of the ^{15}N-labeled Dvl PDZ domain (residues 247–341 of mouse Dvl1) was recorded as a reference.
2. Compound was added to the solution PDZ domain and NMR spectrum of the mixture was recorded. If the compound generated perturbations to the PDZ domain, different amounts of compound were added to the PDZ domain and series of NMR spectra were recorded for calculations of compound's dissociation constant (K_D) to the PDZ domain.
3. NMR spectra were processed with NMRpipe (22).
4. NMR spectra were analyzed by using Sparky (23). We monitored whether added compounds generated chemical shift perturbations to the resonances of the Dvl PDZ domain and compared if the chemical shift perturbations were similar to those caused by binding of the Dapper peptide and Fz7 peptide, which was derived from an Fz membrane receptor (1). Similar perturbations would suggest that compound binds to the same binding site as native PDZ domain-binding partners such as Dapper and Fz. Among nine tested compounds, NSC668036 perturbed the PDZ domain similarly as Dapper and Fz suggesting that it binds to the same binding site as these two native peptides (Note 4).
5. Calculating K_D by using HSQC titration spectra: The binding affinities (K_D) of PDZ ligands were calculated using HSQC spectra by following the method described by Worrall et al. (24). The mean chemical-shift perturbation changes caused by the binding of ligands were calculated using Eq. 1. K_D was then calculated using Eqs. 2 and 3 by applying a one-site binding model with corrections for dilutions, where R was the ligand-to-protein molar ratio, P was the protein concentration before titration, C was the ligand stock concentration, and K_D was the dissociation constant. Two-parameter nonlinear least-squares fitting was performed with program Prism (GraphPad Software, La Jolla, CA):

$$\Delta\delta_{avg} = \sqrt{1/2(\Delta\delta N/5)^2 + (\Delta\delta H)^2} \quad (1)$$

$$\Delta\delta_{binding} = (1/2)\Delta\delta_{max}(A - \sqrt{A^2 - 4R}). \quad (2)$$

$$A = 1 + R + (PR + C)K_D / (PC). \quad (3)$$

3.3. Molecular Dynamics Simulation of Verified PDZ Ligand

To further investigate the interaction between the PDZ domain and NSC668036, the AMBER software (25) suite was used to conduct an MD simulation study of the NSC668036–PDZ domain complex.

1. The starting structures of ligand–protein complexes were prepared using the output from the FlexX docking studies. To sample sufficient possible binding modes during the MD simulation, the entire output of the initial FlexX docking results was reexamined. Based on structure alignment, the 30 docked NSC668036 conformers can be grouped into three clusters. Manual inspection of these docking conformers led us to select 10 conformers from the three clusters representing the 30 conformers as starting points for the MD simulations.
2. After neutralization of these complexes with Na^+ or Cl^-, they were dissolved in a periodic rectangular TIP3P water box, with each side 10 Å from the edge of the complex. AM1-BCC charges and parameters from GAFF force field were assigned to NSC668036 by using the Antechamber module (26) in AMBER 8. For protein, ions, and water, parm99 force field (25, 27) was used.
3. Systems were minimized by a 1,000-step steepest descent minimization followed by a 9,000-step conjugated gradient minimization (25, 27).
4. MD simulation was performed by using the sander program in AMBER 8 with a time step of 2 fs and the nonbonded cutoff set to 9.0 Å. Constant volume (NVT) and constant pressure (NPT) ensemble simulations were carried out to equilibrate the system. In detail, a 50 ps NVT simulation was used to increase the temperature from 100 to 300 K; then the 50 ps NPT ensemble was used to adjust the solvent density, and another 100 ps NPT ensemble was used to gradually reduce the harmonic restraints from 5.0 kcal/mol/$Å^2$ to none (25, 27).
5. MD simulations were performed in explicit water for 5 ns after equilibration with the particle mesh Ewald (PME) method (28, 29) using the NPT ensemble. During the production run, snapshots were saved every 5 ps. Other simulation parameters were set to values similar to those described in the work by Gohlke et al. (30).

3.4. Binding Free Energy Calculation

1. The MM-PBSA algorithm (7–9) was then used to calculate the binding free energy of the interaction between the PDZ domain and NSC668036 using Eq. 4 with the mm_pbsa.pl script in AMBER 8, which employs an MM-PBSA approach:

$$\Delta G_{\text{total}} = G^{\text{complex}} - G^{\text{protein}} - G^{\text{ligand}}, \tag{4}$$

where

$$G = H_{\text{gas}} + H_{\text{trans/rot}} + G_{\text{sovlation}} - TS \tag{5}$$

$$G_{\text{so!vation}} = G^{\text{polar}}_{\text{solvation}} + G^{\text{nonpolar}}_{\text{solvation}} \tag{6}$$

$$G^{\text{nonpolar}}_{\text{solvation}} = \gamma A + b, \tag{7}$$

where the gas phase energy, H_{gas}, is the sum of internal (bond, angle, and torsion), van der Waals, and electrostatic energies in the molecular mechanical force field with no cutoff, as calculated by molecular mechanics (31). $H_{\text{trans/rot}}$ is $3RT$ (R being the gas constant) because of six translational and rotational degrees of freedom. The solvation free energy, $G_{\text{solvation}}$, was calculated by using the PB model (8, 30, 31). In PB calculations, the polar solvation energy, $G^{\text{polar}}_{\text{solvation}}$, was obtained by solving the PB equation with Delphi using parse radius, parm94 charges (for the PDZ domain and the Dapper peptide), and AM1-BCC charges (for the compound). The nonpolar contribution was calculated by Eq. 7. In this equation, A is the solvent accessible area calculated by the Molsurf module in Amber 8 and γ (surface tension) and b (a constant) were 0.00542 kcal/mol/Å^2 and 0.92 kcal/mol, respectively.

2. All of the energy terms given above were averaged from 150 snapshots extracted every 20 ps, and the entropy TS was estimated by normal-mode analysis using 15 snapshots extracted every 200 ps during the last 3-ns production run.
3. During the 10 MD simulation runs, the simulation that started with conformer 22 had the lowest and most stable binding free energy, suggesting that this conformer represents the true PDZ domain-bound conformation of NSC668036 in solution.

3.5. Deduction of Pharmacophore Based on a PDZ–Inhibitor Complex Structure and Two Non-binders

A pharmacophore is composed of functional groups essential and necessary for receptor–ligand binding; without those groups, ligands will no longer bind to receptors and lose their activities (32). Pharmacophore-based approaches have been widely used and shown successes in the field of computer-aided drug design (33, 34). In order to identify more PDZ domain inhibitors, we took such an approach, derived the pharmacophore of PDZ ligands, and used it to screen for PDZ domain inhibitors in the ChemDiv database.

1. The complex structure of NSC668036 and the PDZ domain generated by docking and extensive molecular dynamics simulations was used to build a pharmacophore by using LigandScout (Inte:Ligand, Austria). LigandScout extracts 3D pharmacophores based on complex structures (35).
2. The LigandScout pharmacophore model of NSC668036 is consistent with the structural analysis. The carboxyl group contributes three hydrogen-bond acceptors and the isopropyl group contributes a hydrophobic interaction to the pharmacophore.
3. The NSC668036 compound was compared with other compounds that did not bind to the PDZ domain as verified by NMR experiments. By aligning two non-binders, NSC344681 (((((2-amino-3-hydroxybutanoyl) amino)acetyl)amino)acetic acid) and NSC119132 (methyl 3-((2-amino-3-(aminooxy) propanoyl)amino)-2-((aminooxy) methyl)-3-oxopropanoate) against NSC668036, the differences among these structures that might render NSC344681 and NSC119132 inactive were identified. NSC334681 does not have a hydrophobic group at the 2-position; and NSC119132 is an ester instead of a free acid. This finding strongly suggested that both the carboxyl group and the hydrophobic group next to it were important to the binding and might compose the pharmacophore of PDZ ligands since compounds without them did not bind to the protein.

3.6. Pharmacophore-Based Virtual Screening

1. Based on the pharmacophore proposed above, a 2D search to retrieve all the compounds with 2-(3-methylbutanoic acid) groups in the ChemDiv database by using UNITY in SYBYL® (Tripos, Inc.) was carried out and the search returned 116 hits.
2. To reduce the number of compounds needed to test experimentally and to select ligands with higher potentials to bind to the PDZ domain, the hits were further filtered by using FlexX docking and "Cscore" ranking following the procedure described in Subheading 2.3.1. After manually inspecting top-ranked compounds, 15 compounds were chosen for further examinations based on the following four criteria: (a) the compounds were docked into the designated binding site, (b) their docking conformations were complementary to those of the PDZ domain, and (c) the docked compounds formed hydrogen-bonds with the βA–βB loop as well as (d) additional hydrogen bonds with the PDZ domain.

3.7. Build SAR Models of PDZ Domain Inhibitors

1. The interactions between the 15 compounds and the Dvl PDZ domain were examined experimentally by using NMR chemical shift perturbation experiments as described in Subheading 2.3.2. All the 15 compounds bound to the PDZ domain at the same binding site as the Fz7 peptide.

2. The above-verified compounds were then ranked by their K_D to the PDZ domain, which were calculated by using chemical-shift perturbation changes as described in Subheading 2.3.2.
3. The complex structures of these compounds with the PDZ domain were modeled. The FlexX docking models of the 15 PDZ ligands were refined by using Glide (Schrödinger Inc.). During Glide docking, compounds' amide bonds were kept rigid; hydrogen-bond pharmacophores were designed on the protein to induce ligands to form hydrogen-bonds with the βA–βB loop and the βB strand of the PDZ domain. Other than fixed amide bonds and H-bond pharmacophores, default docking parameters were used. All ligand binding poses generated by Glide have reasonable Glide scores, suggesting that they are likely very close to the true binding modes. For example, the Glide score of compound ChemDiv 5435-0027 is –7.57. According to Schrödinger Inc., low-micromolar inhibitors should have scores around –7. Glide 2.5 predicted binding affinities of a set of 125 crystallized complexes with an RMSD of 2.2 kcal/mol against experimental data (36). Based on the facts that compound ChemDiv 5435-0027 binds to the PDZ domain with a moderate binding affinity and its glide score fits with the experimental data, it is likely that this docking conformation is close to the true binding mode. This conformation was then used to generate complex structures for compounds ChemDiv 2509-0036 and 2509-0040 which had different docking poses than those of the rest of 13 compounds. Complex structures of compounds ChemDiv 2509-0036 and 2509-0040 were modeled by superimposing them onto docked compound ChemDiv 5435-0027 followed by ligand minimization in the ligand-binding pocket of the PDZ domain with LigandScout.
4. Structure-based pharmacophore models of these compounds were built using LigandScout as in Subheading 2.3.5.
5. These 15 compounds are similar in structure but have considerable differences in binding affinities, and thus provide an opportunity to study their SAR to gain insights into the molecular determinants of PDZ–ligand binding which might facilitate further ligand optimization. These 15 compounds have similar scaffolds except that some of them have an extra double bond in their scaffold. According to their scaffolds, these compounds can be grouped into two classes and by comparing their structures and binding strengths in the context of docking complex structures and complex structure-based pharmacophore model, the SAR models of these 15 compounds can be built. In turn, the SAR models can be used to guide furthering optimization of PDZ ligands.

4. Notes

1. Using the Structure of a Homologous Protein in VLS and Docking
   To build the search query for the virtual screening stage, we used the crystal structure of the PDZ domain of Xenopus Dvl bound with the Dapper peptide (6) instead of the NMR solution structure of the apo-PDZ domain of mouse Dvl (1). The two PDZ domains share a high degree of homology, especially around the peptide-binding sites; near the binding sites, there is only a single amino acid difference between the two PDZ domains (Glu323 in the PDZ domain of mDvl1 vs. Asp326 in the PDZ domain of Xenopus Dvl), and the side chain of this residue points away from the peptide-binding cleft. The peptide-binding cavity of the domain is smaller in the apo form of the solution structure than in the crystal structure of the Dapper-bound PDZ domain of Xenopus Dvl. This difference is consistent with the classic "induce-and-fit" mechanism, in which, upon the binding of a peptide or a small organic molecule, the binding sites in the PDZ domain undergo conformational change to accommodate the bound ligand. However, this flexibility cannot be fully explored through a UNITY search and the FlexX docking protocols. Therefore, although the PDZ domain of mouse Dvl was used in the experimental studies, the crystal structure of the PDZ domain of Xenopus Dvl provides a better template for the virtual screening steps. Indeed, the binding free energies calculated from MD simulation of the PDZ domain–NSC668036 and PDZ domain–Dapper peptide complexes fit well with the experimental binding data.
2. Docking Validation
   As a control, we also docked the Dapper peptide into the PDZ domain using FlexX. Under this condition, the docked Dapper peptide had a conformation similar to that found in the crystal structure of the complex with a backbone RMSD of 2.04 Å. In particular, the backbone RMSD for the last six C-terminal amino acids is 1.22 Å, indicating that the docking procedure we used was able to dock ligand into the binding site of the PDZ domain with reasonable accuracy.
3. Compound Handling
   To test compounds using NMR experiments, they should be dissolved in the same buffer as in which the protein was dissolved. pH should be adjusted back to 7.5 if changed to ensure that perturbations on the PDZ domain would only come from the compounds but not from changes in the condition. For compounds with small solubility in the protein buffer, small amount of DMSO was added so that there would be no more than 5%

of DMSO in the final mixture of the PDZ domain with the compound. NMR experiments showed that <5% DMSO does not change the spectra of the PDZ domain. Compound solutions were stored in the freezer at −20°C to avoid hydrolysis and/or other reactions that might change compound identities.

4. Test Selectivity of Identified Ligands
We tested two other PDZ domains: the first PDZ domain of PSD-95, PSD95a (37) (PDB entries 1IU0 and 1IU2), which belongs to the class I PDZ domains, and the PDZ7 domain of the glutamate receptor-interacting protein (38) (PDB entry 1M5Z), a member of the class II PDZ domains. NSC668036 binds to both of these PDZ domains extremely weakly.

References

1. Wong H-C, et al. Direct binding of the PDZ domain of Dishevelled to a conserved internal sequence in the c-terminal region of Frizzled. Mol Cell. 2003;12:1251–60.
2. Barker N, Clevers H. Mining the Wnt pathway for cancer therapeutics. Nat Rev Drug Discov. 2006;5:997–1014.
3. Grandy D, et al. Discovery and characterization of a small molecule inhibitor of the PDZ domain of dishevelled. J Biol Chem. 2009; 284:16256–63.
4. Lee HJ, et al. Sulindac inhibits canonical Wnt signaling by blocking the PDZ domain of the protein Dishevelled. Angew Chem Int Ed Engl. 2009;48:6448–52.
5. Lee HJ, Zheng JJ. PDZ domains and their binding partners: structure, specificity, and modification. Cell Commun Signal. 2010;8:8.
6. Cheyette BNR, et al. Dapper, a Dishevelled-associated antagonist of b-Catenin and JNK signaling, is required for Notochord formation. Dev Cell. 2002;2:449–61.
7. Gohlke H, Case DA. Converging free energy estimates: MM-PB(GB)SA studies on the protein–protein complex Ras–Raf. J Comput Chem. 2004;25:238–50.
8. Wang J, et al. Use of MM-PBSA in reproducing the binding free energies to HIV-1 RT of TIBO derivatives and predicting the binding mode to HIV-1 RT of efavirenz by docking and MM-PBSA. J Am Chem Soc. 2001;123:5221–30.
9. Wang W, et al. BIOMOLECULAR SIMULATIONS: Recent developments in force fields, simulations of enzyme catalysis, protein-ligand, protein-protein, and protein-nucleic acid noncovalent interactions. Annu Rev Biophys Biomol Struct. 2001;30:211–43.
10. Burbaum JJ, Sigal NH. New technologies for high-throughput screening. Curr Opin Chem Biol. 1997;1:72–8.
11. Liu B, et al. Technological advances in high-throughput screening. Am J Pharmacogenomics. 2004;4:263–76.
12. Bleicher KH, et al. Hit and lead generation: beyond high-throughput screening. Nat Rev Drug Discov. 2003;2:369–78.
13. Keseru GM, Makara GM. Hit discovery and hit-to-lead approaches. Drug Discov Today. 2006;11:741–8.
14. Irwin JJ, Shoichet BK. ZINC—a free database of commercially available compounds for virtual screening. J Chem Inf Model. 2004; 45:177–82.
15. Hann MM, Oprea TI. Pursuing the leadlikeness concept in pharmaceutical research. Curr Opin Chem Biol. 2004;8:255–63.
16. Shan J, et al. Identification of a specific inhibitor of the Dishevelled PDZ domain. Biochemistry. 2005;44:15495–503.
17. London TBC, et al. Interaction between the internal motif KTXXXI of Idax and mDvl PDZ domain. Biochem Biophys Res Commun. 2004;322:326–32.
18. Hurst T. Flexible 3D searching: the directed tweak technique. J Chem Inf Comput Sci. 1994;34:190–6.
19. Rarey M, et al. A fast flexible docking method using an incremental construction algorithm. J Mol Biol. 1996;261:470–89.
20. Clark RD, et al. Consensus scoring for ligand/protein interactions. J Mol Graph Model. 2002;20:281–95.
21. Zheng J, et al. Identification of the binding site for acidic phospholipids on the PH domain of

dynamin: Implications for stimulation of GTPase activity. J Mol Biol. 1996;255:14–21.

22. Delaglio F, et al. NMRPipe: a multidimensional spectral processing system based on UNIX pipes. J Biomol NMR. 1995;6:277–93.
23. Goddard TD, Kenller DG SPARKY 3. University of California, San Francisco 2008.
24. Worrall JAR, et al. Transient protein interactions studied by NMR spectroscopy: the case of cytochrome c and adrenodoxin. Biochemistry. 2003;42:7068–76.
25. Case DA, et al. AMBER 8. La Jolla: Scripps Research Institute; 2004.
26. Wang J, et al. Development and testing of a general amber force field. J Comput Chem. 2004;25:1157–74.
27. Cornell WD, et al. A second generation force field for the simulation of proteins, nucleic acids, and organic molecules. J Am Chem Soc. 1995;117:5179–97.
28. Cerutti DS, et al. Staggered Mesh Ewald: an extension of the Smooth Particle-Mesh Ewald method adding great versatility. J Chem Theory Comput. 2009;5:2322.
29. Simmerling C, et al. Combined locally enhanced sampling and Particle Mesh Ewald as a strategy to locate the experimental structure of a nonhelical nucleic acid. J Am Chem Soc. 1998;120:7149–55.
30. Gohlke H, et al. Insights into protein-protein binding by binding free energy calculation and free energy decomposition for the Ras-Raf and Ras-RalGDS complexes. J Mol Biol. 2003; 330:891–913.
31. Kollman PA, et al. Calculating structures and free energies of complex molecules: combining molecular mechanics and continuum models. Acc Chem Res. 2000;33:889–97.
32. Gund P. Three-dimensional pharmacophoric pattern searching. Prog Mol Subcell Biol. 1977;5:17.
33. Guner OF. Pharmacophore perception, development, and use in drug design. La Jolla: International University Line; 2000.
34. Langer T, Hoffmann RD. Pharmacophores and pharmacophore searches. Wiley: Weinheim; 2006.
35. Wolber G, Langer T. LigandScout: 3-D pharmacophores derived from protein-bound ligands and their use as virtual screening filters. J Chem Inf Model. 2004;45:160–9.
36. Friesner RA, et al. Glide: a new approach for rapid, accurate docking and scoring. 1. Method and assessment of docking accuracy. J Med Chem. 2004;47:1739–49.
37. Long J-F, et al. Supramodular structure and synergistic target binding of the N-terminal tandem PDZ domains of PSD-95. J Mol Biol. 2003;327:203–14.
38. Feng W, et al. PDZ7 of glutamate receptor interacting protein binds to its target via a novel hydrophobic surface area. J Biol Chem. 2002; 277:41140–6.

Chapter 3

Rational Design of Rho GTPase-Targeting Inhibitors

Xun Shang and Yi Zheng

Abstract

Rho GTPases have been implicated in diverse cellular functions and are potential therapeutic targets in inflammation, cancer, and neurologic diseases. Virtual screening of compounds that fit into surface grooves of RhoA known to be critical for guanine nucleotide exchange factor (GEF) interaction produced chemical candidates with minimized docking energy. Subsequent screening for inhibitory activity of RhoA binding to the Rho-GEF, LARG, identified a Rho-specific inhibitor as a lead compound capable of blocking RhoA–LARG interaction and RhoA activation by LARG specifically and dose dependently. A microscale thermophoresis analysis was applied to directly quantify the binding interaction of the lead inhibitor with RhoA target. The lead inhibitor highlights the principle that rational targeting of subfamily members of Rho GTPases is feasible and potentially useful in future drug design effort.

Key words: Rho GTPases, RhoA, Signaling, Small molecule, Inhibitor, Rational drug design, Targeting

1. Introduction

Rho GTPases are critical signaling molecules that regulate cytoskeleton organization, gene expression, cell cycle progression, cell motility, and other cellular processes (1–3). They have been implicated in multiple aspects of inflammation, cancer development, and neurologic disorders (4, 5). Understanding the physiological and pathological function of each Rho GTPase and devising novel approaches targeting members of Rho family may have significant biological and therapeutic values (6, 7).

Classic Rho GTPases including RhoA cycle between the GTP-bound, active and the GDP-bound, inactive states. They can be activated through the interaction with the Dbl family guanine nucleotide exchange factors (GEFs) that catalyze their GTP–GDP exchange and connect them to the diverse stimuli from upstream cell surface receptors such as the G protein-coupled receptors,

Yi Zheng (ed.), *Rational Drug Design: Methods and Protocols*, Methods in Molecular Biology, vol. 928,
DOI 10.1007/978-1-62703-008-3_3, © Springer Science+Business Media New York 2012

growth factor receptors, cytokine receptors, and adhesion receptors. In turn, the GTP-bound, activated Rho GTPases turn on an array of cellular responses by engaging specific downstream effectors. Multiple Rho GTPase family members have been suggested as potential targets in a variety of diseases (4, 5).

Extensive efforts have been devoted to the design of inhibitors targeting regulators (e.g., farnesyl transferases (8)), Rho GTPases (e.g., by bacteria toxins (9) or small molecules (10, 11)), and downstream effectors (e.g., ROCK and SRF, refs. 12, 13). However, no clinical utility has been achieved to date due to their nonspecific nature.

Small GTPases, like Ras, are considered undruggable because these small GTPases have a globular structure with limited druggable surface areas, a structure property that makes it difficult to devise means for high-affinity binding by small molecule compounds (14). The previously identified lead Rac1-specific small molecule inhibitor, NSC23766, that has a binding affinity to Rac1 in the micromolar range has been difficult to improve (6). Thus, one key hurdle in the design of effective chemicals affecting Rho GTPase activities relates to the potency and specificity of the small molecules that show activity with individual members of Rho GTPases and/or GTPase-regulated pathways. Within this Rho GTPase regulatory signaling module, several potential sites of intervention strategies have been proposed for effective suppression of Rho GTPase activities (7). Among these potential intervention strategies, targeting the GEF–Rho GTPase interactive surface that is essential for the activation of GDP/GTP exchange reaction has been successfully used in the rational design of Rac1 inhibitor, NSC23766, that specifically binds to the surface groove of Rac1 required for GEF interaction and effectively inhibits Rac1 activity in diverse physiological and pathological systems (11).

Extensive structural studies of RhoA GTPase interaction with its activator GEFs (15) led us to hypothesize that small molecules bound to the surface of RhoA involved in recognition by its GEFs could similarly inhibit RhoA activity and consequent downstream signaling. By structure-based rational design, we have identified a lead small molecule inhibitor that specifically targets RhoA and closely related RhoB and RhoC. This inhibitor specifically inhibits the binding of RhoA to LARG (a RhoGEF protein) and effectively suppresses the RhoA activity and signaling in cells.

2. Materials

2.1. Virtual Screening

1. YASARA software.
2. OpenEye Scientific Software.

2.2. In Vitro Complex Formation Assay

1. Glutathione-agarose beads.
2. 6× His tag® antibody.
3. Recombinant human LARG (residues 765–1,138) containing the DH-PH catalytic module was expressed in *Escherichia coli* BL21 (DE3) strain as N-terminal His-6-tagged fusion proteins by using the pET expression system. Human full-length RhoA (residues 1–193) and human full-length Cdc42 (1–191) were expressed in *E. coli* DH5α strain as GST fusions by using the pGEX-KG vector. The N-terminal tagged GST or $(His)_6$ fusion proteins were purified by glutathione- or Ni^{2+}-agarose affinity chromatography.
4. Compounds G01, G03, G04, G05, G10, and G11 (see Note 1).
5. Lysis buffer: 20 mM Tris–HCl pH 7.6, 100 mM NaCl, 10 mM $MgCl_2$, 0.5% Triton X-100, 2 mM PMSF, 10 μg/mL leupeptin, 10 μg/mL Aprotinin, 0.5 mM DTT (see Note 2).
6. Binding buffer: 20 mM Tris–HCl (pH 7.6), 100 mM NaCl, 1% BSA, 1% Triton X-100, 1 mM $MgCl_2$ (see Note 3).
7. Washing buffer: 20 mM Tris–HCl (pH 7.6), 100 mM NaCl, 1 mM $MgCl_2$.

2.2.1. Microscale Thermophoresis

1. Monolith NT.115 Instrument.
2. Human full-length RhoA (residues 1–193) was expressed in *E. coli* DH5α strain as GST fusions by using the pGEX-KG vector.
3. 50 mM compound G04, a lead RhoA binding small molecule inhibitor (see Fig. 1 and Note 4).
4. Monolith NT™ Protein Labeling Kits (including 0.05 mg NT-647 dye Spin Column A, Buffer Exchange Gravity Flow Column B, Purification Adapter 15 ml Falcons Labeling Buffer).

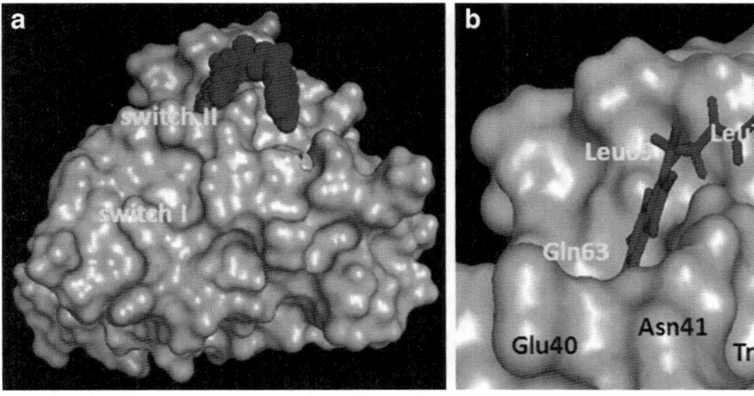

Fig. 1. A simulated docking model of G04 on RhoA surface. The view of the binding pocket of RhoA bound to G04 (*upper*). The view of the predicted structural contacts of G04 in the binding pocket around Trp58 of RhoA (*lower*).

5. Exchange buffer: 50 mM Hepes, 50 mM NaCl, 0.01% Tween 20, and 2 mM $MgCl_2$.
6. 100% DMSO.
7. Monolith NT™ Capillary.

3. Methods

3.1. Virtual Screening

1. The atomic interactions between RhoA and LARG were obtained from PDB ID 1X86 (16). Redundant objects were removed leaving chain A (LARG) and chain B (RhoA).
2. The complex was repaired by adding missing atoms, assigning correct bond orders and resonance, and predicting p*K*a values of exposed residues at pH 7.4 and 0.9% NaCl (17).
3. One iteration of backbone-restrained energy minimization was followed by full energy minimization in explicit solvent using the AMBER99 force field and the Particle Mesh Ewald method for treatment of long-range electrostatics (cutoff 7.2 A°) using YASARA (18).
4. After the protein complex was repaired, LARG was removed and the interaction surface with RhoA examined in order to identify reasonable small molecule binding sites. This is a qualitative process that involves looking for concave surfaces with relatively low thermodynamic motion (low beta factor), and a relatively high density of protein:protein interactions.
5. The concave surface to either side RhoA Trp58 meets these criteria and was chosen as the binding site for virtual screening (Fig. 1).
6. The *in silico* compound library was derived from the ZINC purchasable (International Zinc Association) compound set.
7. Single molecules were used to create conformers using OMEGA at default settings (OpenEye Scientific Software). The conformer library was screened across the W58 binding site using FRED, rigid ligand and receptor, default scoring (OpenEye Scientific Software), in parallel mode on an SGI Altix 4700 computer.
8. The top 1,000 hits were submitted for flexible-ligand, rigid-receptor docking and energy minimization using Molegro, default scoring (Molegro ApS).
9. The top 100 compounds were selected and the top 49 were obtained for testing.

3.2. In Vitro Complex Formation Assay

The interaction between RhoA and LARG can be examined by in vitro complex formation pull-down assay. The LARG protein

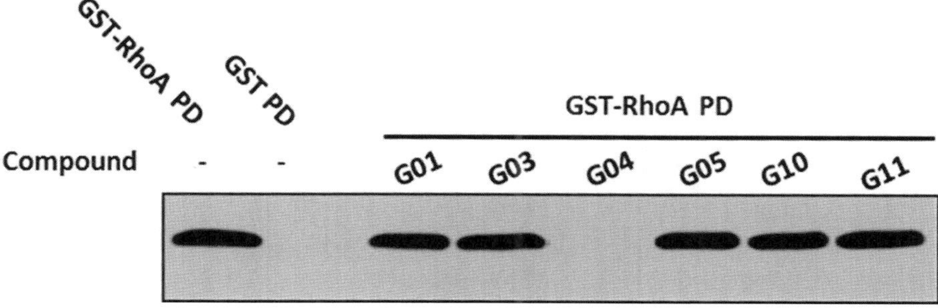

Fig. 2. The inhibitory effect of a panel of compounds predicted by virtual screening on RhoA interaction with LARG DH-PH module was tested in a complex formation assay. $(His)_6$-tagged LARG (1 μg) was incubated with GST alone or GST-RhoA (1 μg) immobilized on glutathione agarose beads in the presence or absence of the 1 mM indicated compounds. After an incubation at 4 °C for 1 h, the beads associated $(His)_6$-LARG were detected by anti-His Western blotting.

binds to the RhoA protein conjugated by glutathione-agarose beads and can be collected by centrifugation and detected by the antibodies (Fig. 2).

1. 1 μg of $(His)_6$-tagged LARG DH-PH was incubated with 1 μg of EDTA-treated (1 mM) GST-fused RhoA in binding buffer and 15 μl of suspended glutathione-agarose beads (see Note 5).
2. Compound G04 or other chemicals were added in the incubation buffer at 1 mM concentrations (see Note 6).
3. After incubation at 4 °C for 1 h under constant agitation, collect the beads by centrifugation at 650 × *g* for 2 min. Discard the supernatants. Wash the beads with 1 mL ice-cold washing buffer for three times.
4. Analyze the beads by SDS-PAGE (4–15% gradient gel). The amount of $(His)_6$-tagged protein co-precipitated with the GST-fusion-bound beads was detected by anti-His Western blotting.

3.3. Microscale Thermophoresis

Microscale Thermophoresis is a powerful new technology. It measures changes in the hydration shell of molecules and allows to quantify enzyme kinetics or to measure biomolecule interactions under close-to-native conditions. Many biochemical processes relating to the interaction, size, stability, or conformation of biomolecules and biochemical complexes can be measured with high sensitivity. The Microscale Thermophoresis Technology can detect binding events which are typically difficult to access, like the binding of small molecules or single ions to proteins (19). Furthermore, it can be used to study the activity of molecules including enzymatic modifications of substrates or competitive binding of inhibitors and substrates to an enzyme. In this study we utilized this technique to quantify direct binding of our lead inhibitor to its target, RhoA protein (Fig. 3).

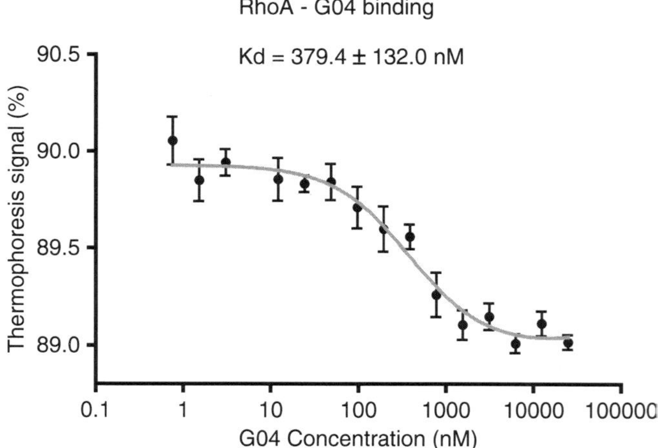

Fig. 3. Microscale thermophoresis analysis of G04 to RhoA. Purified RhoA was first labeled with Alexa-647 fluorescence dye. G04 was titrated between 0.76 and 250,000 nM to a constant amount of labeled RhoA (100 nM). The reaction was performed in 50 mM Hepes, 50 mM NaCl, 0.01% Tween 20, and 2 mM $MgCl_2$. The samples were incubated at room temperature for 1 h before the measurements. The average K_d values ±SD were calculated from three replicates. All the data are representative of three independent experiments.

3.3.1. Buffer Exchange (See Note 7)

1. Add 1.7 ml distilled water to the vials containing the labeling buffer salt.
2. Invert Spin Column-A to suspend slurry.
3. Twist off bottom (twist slightly in both directions).
4. Loose the cap of the column.
5. Place column in 1.5–2 ml microcentrifuge collection tube.
6. Centrifuge at 1,500 × *g* for 1 min to remove excess liquid.
7. Add 300 μl of labeling buffer and repeat step 5 (repeat three times).
8. Place 40–100 μl protein in the center of the resin (see Note 8).
9. Place the sample in a new microcentrifuge collection tube and centrifuge at 1,500 × *g* for 2 min (see Note 9).

3.3.2. RhoA Protein Labeling

1. Adjust RhoA concentration to 2–20 μM using the Labeling Buffer (maximum of 250 μl protein solution can be used per labeling procedure, see Note 10). For Calculation of the molarity of your sample use the following equation: *c*(Protein) [Mol] = *c*(Protein) [mg/ml]/MW (Protein) [Da].
2. Add 50 μl of 100% DMSO to the solid fluorescent dye (i.e., approximately 1.3 mM solution). Mix the dye thoroughly.
3. Adjust the concentration of the dye to two- to threefold concentration of the RhoA protein by using the labeling buffer.
4. Mix Protein and Dye in a 1:1 ratio (500 μl maximum volume per gravity flow column-B).
5. Incubate for 30 min at room temperature in the dark.

3.3.3. Protein Purification

For optimal results using Microscale Thermophoresis, unreacted "free" dye needs to be eliminated. The purity can be monitored by measuring the ratio of protein to dye (e.g., spectroscopically by measuring absorption at 280 nm for protein and 650 nm for the dye, Molar absorbance: 250,000 M^{-1}/cm) after cleanup procedure.

1. Remove the top cap of Column-B and pour off the column storage solution.
2. Remove the bottom cap and place in a 15 ml tube (Adapter supplied).
3. Equilibrate and wash Column B by adding 8 ml of your final protein storage/analysis buffer of choice (Flow through by gravity flow).
4. Add maximum 500 μl of labeling reaction to the center of Column B. Let sample enter the bed completely (when using less than 500 μl, adjust the volume to 500 μl after the sample has entered the bed by using your final storage/analysis buffer; see Note 11).
5. Discard the flow through.
6. Place in a new 15 ml collection tube.
7. Add 600 μl of storage/analysis buffer and collect the elute. The first few μl will not contain protein and can be discarded.
8. Measure the ratio of protein/dye and aliquot/store your protein spectroscopically.

3.3.4. Experiment Preparation

1. Prepare the labeled RhoA: Ideally the ratio of label to protein (DEG, degree of labeling) is 1:1 or less. Unbound label has to be eliminated from the labeling mixture, completely. The concentration of the protein should be close to the expected KD or less and should yield a range of 80–1,500 fluorescence counts (typically equivalent to 1–200 nM) when measured with the Monolith NT.115.
2. Prepare the unlabeled binding partner-G04. Prepare a 16-step dilution series around the expected KD, starting from a concentration of 20-fold the expected KD down to 0.02-fold the expected KD (see Note 12). Prepare the samples with a two-fold higher concentration, since mixing volume ratio of 1:1 decreases the concentration by a factor of 2. Use sufficiently small microreaction tubes to avoid evaporation of the sample (e.g., with a volume <100 μl when preparing 10–20 μl of sample; see Note 13).

3.3.5. Binding Reaction

1. Prepare a serial dilution of G04 and fill 10 μl of each sample of the dilution series into 16 small microreaction tubes (see Note 14). Fill 10 μl of a 2× stock solution of G04 into the same 16 tubes and mix very well. Incubate the samples in the dark.

Incubation time and temperature can differ between different molecules and should be tested in a pilot experiment by the researcher. In most cases 30–60-min incubation at room temperature is sufficient.

2. Transfer the binding reaction into Monolith NT™ Capillaries.

3.3.6. Sample Loading

1. Stick the Monolith NT™ Capillary horizontally into the reaction tube to aspirate the sample (see Note 15). It is recommended to stop sucking in the liquid when the capillary is filled to 2/3 of its length.
2. Position the liquid in the center of the Monolith NT™ Capillary by holding it vertically and shake it after aspirating the sample. Leave some air at both ends of the capillary.
3. Seal the ends with wax.
4. Put the capillaries in the slots on the sample tray (see Note 16).
5. Place the tray in the instrument by pushing it into the try slot of the instrument as far as possible. The Monolith NT.115 instrument scans the tray and automatically determines the position of the capillaries on the tray.

3.3.7. Run the Monolith NT.115 Instrument

Please see the user manual of Nano Temper, Germany.

4. Notes

1. All the compounds are dissolved in DMSO and the stocking concentration is 50 mM.
2. The lysis buffer is to lysate pET or pGEX-KG plasmids transformed bacterial.
3. 1% BSA was added to the lysis buffer to reduce the stickiness of the LARG protein to the wall of the test tube.
4. G04, a RhoA targeting small molecule inhibitor shown in Fig. 1, is dissolved in DMSO.
5. Previous reports show that LARG DH-PH domain binds to RhoA (20).
6. This is final concentration.
7. The labeling step requires that the protein is dissolved in a suitable labeling buffer at the correct pH. The purified protein should be in a buffer that does not contain primary amines (e.g., ammonium ions, Tris, glycine, ethanolamine, triethylamine, glutathione) or imidazole. All of these substances will significantly reduce protein labeling. Also, partially purified

protein samples, or protein samples containing carriers like BSA, will not be labeled properly and should not be used.

8. Be careful not to disturb the resin or allow the sample to flow around the resin bed.
9. Do not exceed recommended centrifuge time or speeds.
10. Make sure that the protein concentration is sufficiently high and in the range of 2–20 μM. Lower concentrations may result in loss of coupling efficiency.
11. To prevent a contamination of your sample with uncoupled dye, we recommend to load the sample directly in the center of the resin bed (Column-A and Column-B). Do not blow out the tips and avoid the contact of the tip with the walls of the column.
12. The concentration of the labeled molecule should be in the range of the expected KD or less.
13. The concentration of the titrated compound should vary from 20× to 0.02× the expected KD.
14. Prepare at least 20 μl of sample per capillary to avoid pipetting error and changes in concentration due to sample evaporation. You will need approximately 5 μl to fill a capillary.
15. Do not touch the capillary in the middle (the optical measurement will be performed at this position).
16. Typically the highest concentration is placed in the front of the tray. This position is denoted as “1” in the Monolith software.

References

1. Etienne-Manneville S, Hall A (2002) Rho GTPases in cell biology. Nature 420:629–635
2. Ridley AJ (2001) Rho family proteins: coordinating cell responses. Trends Cell Biol 11:471–477
3. Zohn IM, Campbell SL, Khosravi-Far R, Rossman KL, Der CL (1998) Rho family proteins and Ras transformation: the RHOad less traveled gets congested. Oncogene 17:1415–1438
4. Sahai E, Marshall CJ (2002) RHO-GTPases and cancer. Nat Rev Cancer 2:133–142
5. Boettner B, Van Aelst L (2002) The role of Rho GTPases in disease development. Gene 286:155–174
6. Nassar N, Cancelas J, Zheng J, Williams DA, Zheng Y (2006) Structure-function based design of small molecule inhibitors targeting Rho family GTPases. Curr Top Med Chem 6:1109–1116
7. Marchioni F, Zheng Y (2009) Targeting rho GTPases by peptidic structures. Curr Pharm Des 15:2481–2487
8. Sebti SM, Der CJ (2003) Opinion: searching for the elusive targets of farnesyltransferase inhibitors. Nat Rev Cancer 3:945–951
9. Genth H, Dreger SC, Huelsenbeck J, Just I (2008) Clostridium difficile toxins: more than mere inhibitors of Rho proteins. Int J Biochem Cell Biol 40:592–597
10. Gao Y, Dickerson JB, Guo F, Zheng J, Zheng Y (2004) Rational design and characterization of a Rac GTPase-specific small molecule inhibitor. Proc Natl Acad Sci USA 101:7618–7623
11. Onesto C, Shutes A, Picard V, Schweighoffer F, Der CJ (2008) Characterization of EHT 1864, a novel small molecule inhibitor of Rac family small GTPases. Methods Enzymol 439:111–129
12. Narumiya S, Ishizaki T, Uehata M (2000) Use and properties of ROCK-specific inhibitor Y-27632. Methods Enzymol 325:273–284
13. Evelyn CR et al (2010) Design, synthesis and prostate cancer cell-based studies of analogs of the Rho/MKL1 transcriptional pathway

inhibitor, CCG-1423. Bioorg Med Chem Lett 20:665–672

14. Vetter IR, Wittinghofer A (2001) The guanine nucleotide-binding switch in three dimensions. Science 294:1299–1304
15. Rossman KL, Der CJ, Sondek J (2005) GEF means go: turning on RHO GTPases with guanine nucleotide-exchange factors. Nat Rev Mol Cell Biol 6:167–180
16. Kristelly R, Gao G, Tesmer JJ (2004) Structural determinants of RhoA binding and nucleotide exchange in leukemia-associated Rho guanine-nucleotide exchange factor. J Biol Chem 279:47352–47362
17. Krieger E, Nielsen JE, Spronk CA, Kollman PA (2006) Fast empirical pKa prediction by Ewald summation. J Mol Graph Model 25:481–486
18. Krieger E, Darden T, Nabuurs S, Finkelstein A, Vriend G (2004) Making optimal use of empirical energy functions: force-field parameterization in crystal space. Proteins 57:678–683
19. Wienken CJ, Baaske P, Rothbauer U, Braun D, Duhr S (2010) Protein-binding assays in biological liquids using microscale thermophoresis. Nat Commun 19:100
20. Fukuhara S, Chikumi H, Gutkind JS (2000) Leukemia-associated Rho guanine nucleotide exchange factor (LARG) links heterotrimeric G proteins of the G(12) family to Rho. FEBS Lett 485:183–188

Chapter 4

Rational Design of Peptide Ligands Against a Glycolipid by NMR Studies

Wenyong Tong*, Tara Sprules*, Kalle Gehring, and H. Uri Saragovi

Abstract

Ganglioside GD2 is a cell surface glycosphingolipid that is targeted clinically for cancer diagnosis, prognosis, and therapy. The conformations of free GD2 and of GD2 bound to anti-GD2 mAb 3F8 were resolved by saturation transfer difference nuclear magnetic resonance and molecular modeling. Then small molecule cyclic peptide ligands that bind to GD2 selectively were designed, and shown to affect GD2-mediated signal transduction. The solution structure of the GD2-bound conformation of the peptide ligands showed an induced-fit binding mechanism. This work furthers the concept of rationally designing ligands for carbohydrate targets; and may be expanded to other clinically relevant gangliosides.

Key words: Saturation transfer difference NMR, Transferred NOE, Antibody–carbohydrate interactions, Peptide–carbohydrate interaction, NMR structure calculations, Molecular modeling, Ganglioside GD2, Peptide mimetics, Cancer

1. Introduction

Nuclear magnetic resonance (NMR) spectroscopy has become a very important tool for drug discovery, not only because of its ability to elucidate molecular structure but also for its great potential to monitor intermolecular interactions at the atomic level. NMR can be used for assessing target druggability, for pharmacophore identification, for hit validation, and for structure-based design (1). NMR can detect interactions in systems that may not be amenable to crystallography.

In particular, techniques such as saturation transfer difference (STD) NMR and transferred nuclear Overhauser effects (trNOEs) are very useful when working with ligands with dissociation constants

* These two authors contributed equally in this chapter.

Yi Zheng (ed.), *Rational Drug Design: Methods and Protocols*, Methods in Molecular Biology, vol. 928, DOI 10.1007/978-1-62703-008-3_4, © Springer Science+Business Media New York 2012

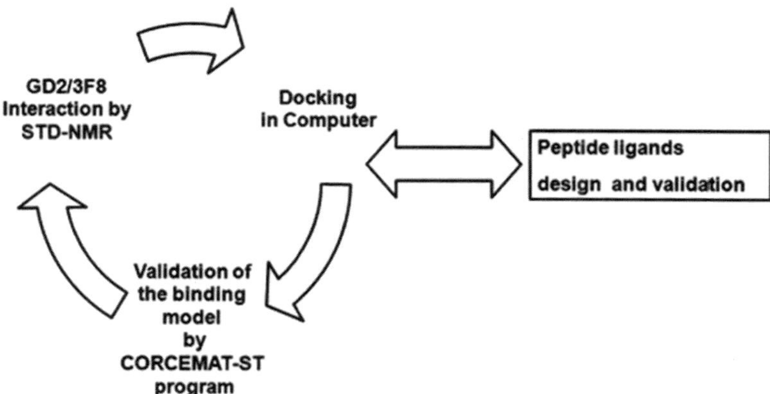

Fig. 1. Structure-based drug design by NMR.

K_D in the range of 10^{-3} to 10^{-8} M. This is because the approach takes advantage of the different NMR relaxation properties of large and small molecules to transfer information from the bound, NMR "invisible" state of the ligand to the free, "visible" state. The STD experiment can identify which functional groups or modules in a ligand are most important for establishing the intermolecular interaction with its target. In the case of monoclonal antibodies (mAbs), it is possible to identify the binding epitope (2).

Transferred NOEs can be used to calculate a three-dimensional (3D) structure of the bound ligand. The NOE experiments provide important information by measuring the average interatomic distances to ~6 Å. Distance information is supplemented by the measurement of torsion angles from through-bond scalar J-couplings. These restraints are input into a molecular modeling algorithm to generate a family of representative structures. Finally, NMR data is complemented by the use of molecular dynamics (MD) simulation. The results can then be applied to rationally design a more potent compound (3). These approaches have been used successfully for protein targets.

Here, a novel example of NMR-based structure design is presented targeting the carbohydrate portion of a cell membrane glycolipid by using STD NMR and trNOEs (Fig. 1) (4). Some glycolipids such as ganglioside GD2 (Fig. 2) are highly expressed in cancer tissues, for example neuroblastomas, small-cell lung carcinomas, and melanomas, but have a very low level of expression in normal tissues. GD2 is thus targeted clinically for diagnosis and immunotherapy using anti-GD2 mAbs, such as IgG3 mAb 3F8 (5). The importance of gangliosides in tumorigenesis, tumor progression, and tumor metastasis may be due to their biological functions that include cell recognition, cell matrix attachment, and cell growth and differentiation (6, 7). Since those cell surface glycolipids can induce the signal transduction cascade, we studied whether NMR-based approaches can generate GD2-selective

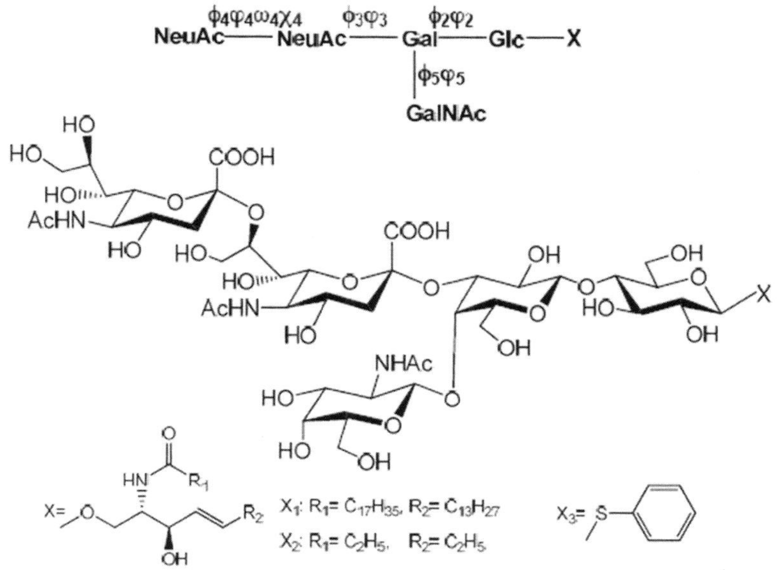

Fig. 2. Structures of ganglioside GD2 and its derivatives. The carbohydrate "head" of GD2 is shown, to which can be attached either a ceramide or a lipid tail (X1); this is herein referred to as ganglioside GD2. The ceramide tail of ganglioside GD2 can be truncated (X2) for the purpose of modeling. X3 is a synthetic water-soluble thiophenyl GD2, lacking the ceramide and lipid tail, as an analog of ganglioside GD2. Residues and torsion angles are shown.

ligands with biological activity. STD NMR spectroscopy was used to study the binding characteristics of anti-GD2 mAb 3F8 to two forms of GD2: (a) intact ganglioside GD2 incorporated in dodecylphosphocholine (DPC) micelles (that forms micelles as a plasma membrane mimic) and (b) a water-soluble phenylthioether-analog of GD2 that lacks ceramide and lipids, the "GD2 carbohydrate head." With the aid of molecular modeling, docking, and theoretical STDs calculated by the CORCEMA-ST program (8), a model of GD2/3F8 was proposed that matched the experimentally determined binding epitope. Then three GD2-selective peptidic ligands, which mimic mAb 3F8, were synthesized by using mAb mimicry designs to keep the β-turn conformation (9). The detergent-like properties of gangliosides (10) were utilized to form lipid raft-mimicking GD2 micelles to study the peptide/carbohydrate interaction by NMR. However, the intensities of observed STDs are primarily dependent upon the longitudinal relaxation rate of the free ligand protons at long saturation times (T1 effects). At the relaxation times used for peptide/GD2 micelle experiments (>2 s) the STD intensities were dominated by T1 effects (11). The conformation of the peptidic GD2 ligands bound to the head group of GD2 was therefore further characterized by trNOE experiments in order to accurately map the binding epitope.

2. Materials

2.1. Chemicals and Regents

1. DPC-d_{38}.
2. Amicon centrifugal filters, 10 kDa cutoff, Millipore.
3. Deuterated water.
4. 50 mM of deuterated phosphate buffer (pD 6.2) made with 70 mg potassium dideuteriumphosphate and 17 mg dipotassium deuterium phosphate in 12.5 ml of 100% D_2O (Ionic strength 0.065).

2.2. Instrumentation

1. Waters 600 HPLC and C-18 semipreparative column.
2. 500 MHz Varian INOVA NMR spectrometer with an H {CN} triple resonance cold probe with Z-gradients and variable temperature control (5–50 °C), and BioPack pulse sequence library.
3. Agilent capLC-MSD Quad.
4. pH meter, with microelectrode that will fit in microfuge tubes or NMR tubes.
5. Heto powerDry 3300 lyophilizer.
6. Nano-drop 1000, UV meter, Wilmington.

2.3. Software for the Analysis of NMR Data

1. Molecular modeling software: ANTECHAMBER (12) and AMBER 9 (13) with the AMBER force field (Glycam_04.dat and GAFF), ARIA 2.2/CNS 1.2 (14), and MMTSB for clustering (15).
2. NMR software: Felix 2004 (Accelrys, Inc), SPARKY (T. D. Goddard and D. G. Kneller, University of California, San Francisco), NMRPipe (16), MestReNova-Lite v5.2.1-3586 (Mestrelab Research SL), CORCEMA-ST program provided by Dr. N. R. Krishna, University of Alabama at Birmingham.
3. Structure presenting software: DiscoveryStudio 1.7 (Accelrys, Inc), PyMOL, and VMD-Xplor.

3. Methods

3.1. Sample Preparation for STD NMR Experiments to Study the Interaction Between Antibody and Carbohydrate

1. Thiophenyl GD2 (see Note 1) (Fig. 2) was synthesized from thiophenyl GD3 (17) using a recombinant β-1,4-*N*-acetylgalactosaminyltransferase (CgtA, construct CJL-30) and an UDP-GlcNAc 4-epimerase (Gne, construct CPG-13) originally cloned from *Campylobacter jejuni* (18, 19). The sample was purified (>98%) by HPLC using a C-18 semipreparative column, and was verified by mass spectrometry (ESI-MS)

and NMR. The measured molecular weight of thiophenyl GD2 corresponded to expected values. Thiophenyl GD2 was dissolved in 50 mM deuterated phosphate buffer (pD 6.2).

2. Mixed micelles were prepared by dissolving ganglioside GD2 (1 mM) and the detergent DPC-d_{38} in a molar ratio of 1:40 in 50 mM deuterated phosphate buffer (pD 6.2); the mixture was exchanged three times with D_2O, with intermediate lyophilization, dried under high vacuum at −56 °C, and finally dissolved in 300 μL of buffer (see Note 2).
3. 3F8 was concentrated using Amicon centrifugal filters and exchanged several times with a 50 mM deuterated phosphate buffer (pD 6.2). The mAb preparations were combined with freeze-dried thiophenyl GD2 or ganglioside GD2 in DPC micelles at molar ratios ranging from 1:25 to 1:200; the final oligosaccharide concentrations were between 0.125 and 1 mM.

3.2. STD NMR Experiments to Study the Interaction Between Antibody and Carbohydrate

3.2.1. NMR Assignments

1D proton spectra, as well as 2D TOCSY, COSY, NOESY, ROESY, ^{13}C HSQC, and ^{13}C HMBC experiments were recorded using the spectrometer's standard pulse sequences for both thiophenyl GD2 (25 °C) and GD2-DPC micelles (35 °C) in the absence of mAb3F8. These experiments give complete ^{1}H and ^{13}C assignments for the two compounds. T1 relaxation rates were measured for thiophenyl GD2 using a standard inversion-recovery pulse sequence (see Note 3).

3.2.2. STD NMR (Fig. 3)

A 1D STD pulse sequence with internal subtraction via phase cycling and a 3-9-19 WATERGATE pulse train for H_2O suppression was employed to record reference and difference STD spectra (see Note 4) (20).

1. 90° high-power proton pulse-widths, the offset for water suppression, and on-resonance saturation (see below) were optimized for each sample.
2. Off-resonance irradiation was set to 34.8 ppm.
3. The irradiation power was $(\gamma/2\pi)B_1 = 65$ Hz, applied through a train of 50 ms eburp1 pulses with a 1 ms delay between the pulses. The relaxation delay was set to 0.1 s in all cases. A 30 ms spin-lock pulse with a strength of $(\gamma/2\pi)B_1 = 1{,}300$ Hz was used to eliminate background antibody/ganglioside signals.
4. Spectra were recorded with 32,000 points, a sweep width of 16 ppm, and 1,024–2,048 transients (see Note 5). A series of spectra with different saturation times (1, 2, 3, and 5 s) and at different ligand:macromolecule ratios were recorded for each system studied. It is important to record an STD spectrum for ligand in the absence of macromolecule to ensure that on-resonance saturation does not perturb any of its signals.
5. Reference spectra were recorded using the same pulse sequence with saturation pulses applied at 34.8 ppm, and no internal subtraction.

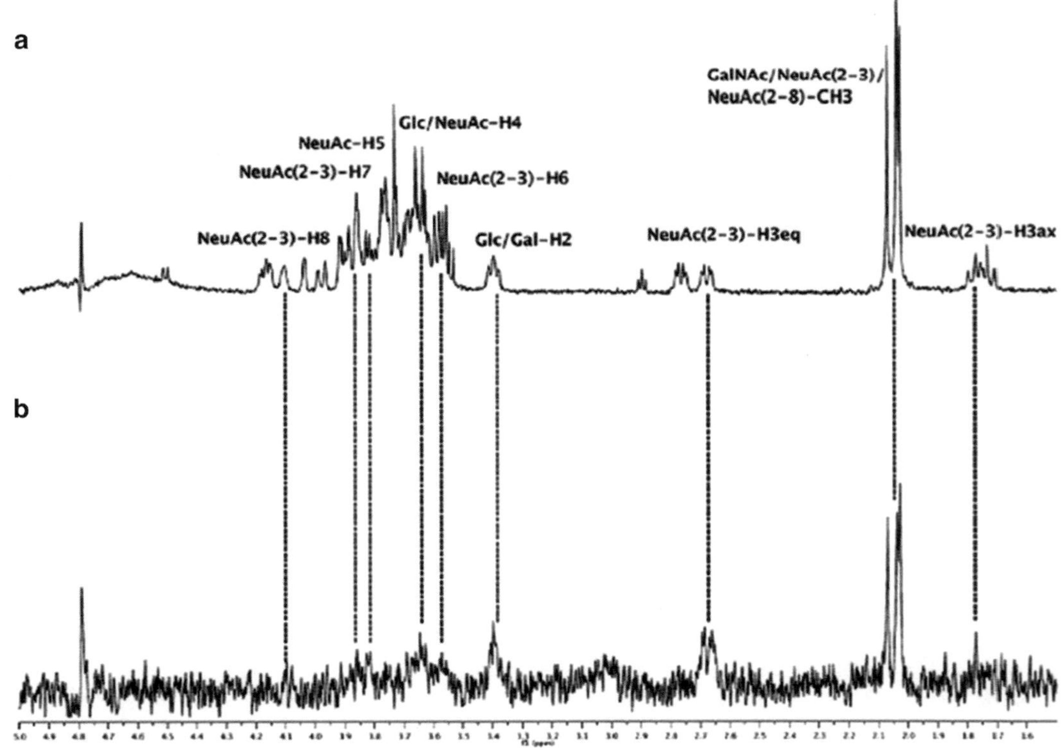

Fig. 3. STD NMR spectra, STD amplification factor, and structure model of thiophenyl GD2 binding to mAb 3F8. (**a**) Reference spectrum of a mixture of thiophenyl GD2 (500 μM) and 3F8 (10 μM binding sites) at a 50:1 ratio. (**b**) STD NMR spectrum of the same sample (T_{sat} = 5 s).

6. Thiophenyl GD2: Reference STD and STD difference spectra (5 s saturation) were recorded for thiophenyl GD2 in the absence of mAb. On-resonance saturation at 0 ppm did not result in any signal in the difference spectrum, indicating that, as expected, the saturation pulse train was far enough away from the ligand resonances, and was not exciting a broader spectral region than it was supposed to. STD difference spectra recorded using the same experimental setup with different saturation times for ligand:mAb ratios from 200:1 to 25:1 showed signals arising from specific protons; therefore saturation was being transferred from mAb 3F8 to the carbohydrate.
7. GD2-DPC micelles: Reference STD and STD difference spectra (5 s saturation) were recorded for GD2-DPC micelles in the absence of mAb. On-resonance saturation at −1 ppm showed some signal at 1.25 ppm, resulting from saturation of the DPC micelles. Upon addition of mAb a series of STD difference spectra with on-resonance saturation at 9.3, 1, −1, and −2 ppm were recorded. Saturation at 9.3 ppm (where only mAb 3F8 signals were expected) and −2 ppm gave very similar results; spectra recorded with on-resonance saturation at 1 and −1 ppm

exhibited strong enhancement of ceramide signals, indicating that the transmitter was too close to the residual proton signals of the DPC micelles. Therefore −2 ppm was used as on-resonance saturation to record a series of STD spectra at different saturation times and ratios of mAb to GD2.

3.3. Structure Calculations of Two Types of GD2

1. Distance restraints derived from NOE or ROE data were produced by the software Felix 2004. The S-M-W bins method, which categorizes connectivities based on distance as strong (2.5 Å), medium (3.5 Å), and weak (6.0 Å), was used to convert the calculated distances to restraints and generate qualitative upper-bound restraints (see Note 6).
2. The intra-residue distance of 1.78 Å between the H3ax and H3eq of *N*-acetyl neuraminic acid was used as a reference for distance calibration (21).
3. In the simulation, the lipid portion of molecular was replaced by a truncated ceramide (Fig. 1). The topology file of the truncated ceramide was produced by ANTECHAMBER.
4. The energy minimization and simulation with distance restraints used the SANDER module of the AMBER 9 program with the AMBER force field (Glycam_04.dat and GAFF) with a dielectric constant of 80. After minimization, the system was heated up to 350 K from 5 K, and then was cooled down to 5 K. Total simulation time was 1 ns.
5. The simulated annealing was repeated (five times) until there was no change in potential energy. This final NMR structure was used as a starting input into 1 ns restrained molecular dynamics (rMD) simulations to produce eight families of conformers according to the combinations of dihedral angles of glycosidic bonds.
6. One conformer from each family was then used as a starting structure for an additional simulated annealing step with NMR refinement.
7. The AMBER 9 suite of programs together with the AMBER force field (Glycam_04.dat and GAFF) was used to perform 10 ns MD simulations for the two types of GD2. Each GD2 conformation was solvated in a truncated octahedron TIP3P water box. Sodium counterions were added to maintain electroneutrality of the system.
8. The system was minimized first, followed by heating from 100 to 303 K over 10 ps in the canonical ensemble (NVT), and by equilibrating to adjust the solvent density under 1 atm pressure over 10 ps in the isothermal isobaric ensemble (NPT) simulation. After an additional 100 ps simulation, a 1 ns production NPT run was obtained with snapshots collected every 1 ps. Langevin dynamics was used for temperature control. Bond lengths involving bonds to hydrogen atoms were constrained by SHAKE.

3.4. Molecular Modeling and Calculation of STD Signals

1. GD2 docking in the putative binding site of Fab fragment of 3F8 was performed with in-house docking software (see Note 7). A box around the six CDRs of the heavy and light chain of 3F8 (see Note 8) was constructed and defined the binding region to be searched.
2. In order to sample various conformations, 12,000 conformers of GD2 were generated from snapshots taken from an MD simulation (12 ns) in explicit water using AMBER ff03 and GAFF force fields starting from the NMR structures. Partial charges of the conformers were calculated by the AM1-BCC method (22).
3. Each of the docked conformers of GD2 was exhaustively docked into the 3F8 Fab and scored. The top-scoring 100 poses were clustered. The most populated cluster was taken to be representative of the true GD2 binding mode in the Fab of 3F8 antibody. The best-scored pose from this cluster was selected to represent the binding model. This pose was subjected to minimization followed by a 5-ns MD simulation.
4. The MD trajectory was used to calculate a predicted binding free energy of the complex using a Solvated Interaction Energy (SIE) approach (23).
5. Theoretical STD values were predicted by the CORCEMA-ST protocol based on the binding mode (Fig. 4) generated from MD simulations for interactions of thiophenyl GD2 and a 3F8 Fab fragment. Spectral densities were calculated only for proton pairs having a distance of 6 Å or less. In the calculations, thiophenyl-GD2 and the 13 amino acid residues within the binding pocket were included (see Note 9).
6. The order parameter S^2 was set to 0.25 for intra-methyl relaxation, while S^2 of 0.85 was used for methyl-X relaxation.
7. The concentration of ligand was 500 μM and the ratio of ligand:protein was 50:1.
8. k_{on} was set to 5×10^7 s^{-1}/M for the best fitting results. The value of the dissociation constant (K_D) for thiophenyl-GD2 that was used in the calculation was 10^{-6} M, calculated from a long-term MD simulation.
9. The final optimized values of the correlation times (τ) were 1.2 and 80 ns for the ligand in the free and bound states, respectively.
10. STD values were calculated as $((I(t)(k) - (I0(k))/I0(k)) \times 100)$, with $I0(k)$ being the intensity of the signal of proton k without saturation transfer at time $t = 0$, and $I(t)(k)$ being the intensity of proton k after saturation transfer during the saturation time t. The NOE R-factor is used to compare theoretical or calculated STD values with the experimental ones. The NOE R-factor is defined as follows (see Note 10) (Table 1).

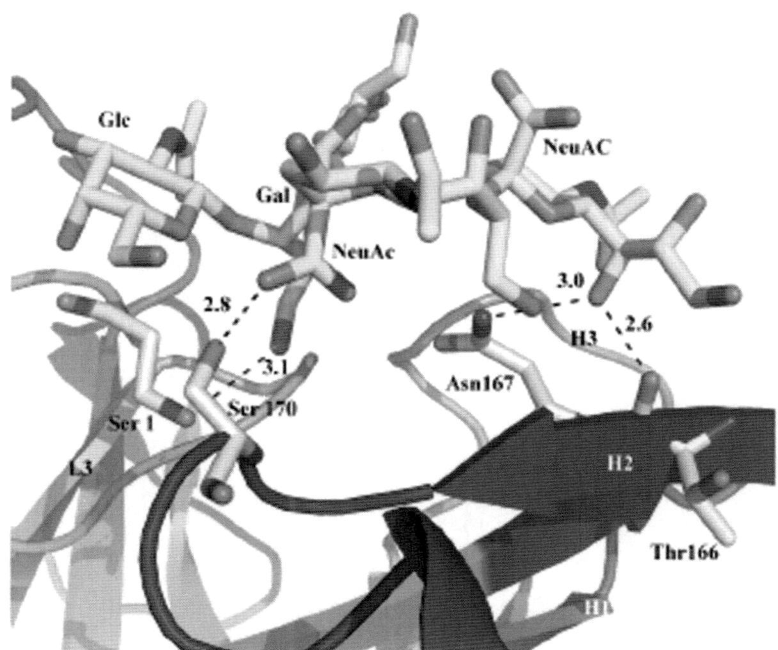

Fig. 4. The predicted binding mode of GD2 in the antigen-binding surface of a Fab fragment of 3F8 antibody. The Fab of 3F8 VL and VH is displayed as a *ribbon* structure. Selected residues of 3F8 are represented as thick capped sticks. GD2 is shown as capped sticks, in (**a**). *Black dashed lines* are H-bonds. Distances are measured between the heavy atoms. Ser1 (framework), Thr 166, Asn 167, and Ser170 (H2) form H-bonds with the GD2 molecule.

Table 1
The NOE *R*-factor equation

$$R\text{ - factor} = \sqrt{\frac{\sum (S_{\text{expt},k} - S_{\text{calk},k})^2}{\sum (S_{\text{expt},k})^2}}$$

3.5. Sample Preparation for STD NMR Experiments to Study the Interaction of Peptides and Glycolipid Micelles

1. Ganglioside GD2 was subjected to three cycles of freeze-drying with D_2O to remove traces of water and to prepare a stock solution with a final concentration of 3 mM in 50 mM deuterated phosphate buffer, pD 6.2. Dilution effects from addition of this GD2 stock solution to peptide ligand solution to make different ratios were minimal (see Note 11).
2. Stock solutions of peptide ligands, 2–3 mM each, were prepared by dissolving them in 50 mM deuterated phosphate buffer (see Note 12). The concentrations of the peptides were determined by UV (Nano-drop 1000) at 280 nM. 1D STD NMR spectra were acquired at 298 K for a variety of peptide-to-GD2 ratios.

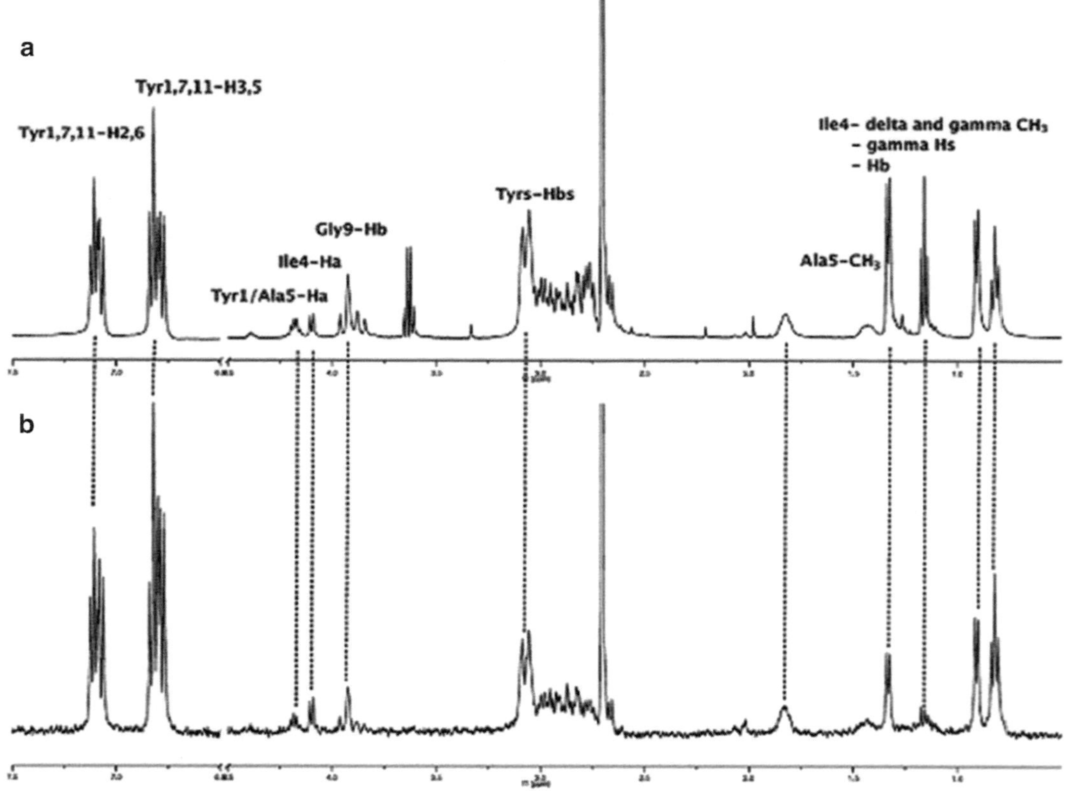

Fig. 5. STD NMR spectra and STD amplification factor for a mixture of ganglioside GD2 micelles and M50B. (**a**) Reference spectrum of a mixture of ganglioside GD2 micelles (200 μM) and peptide ligand M50B (1 mM) in a ratio of 1:5. (**b**) STD NMR spectrum of the same sample at 298 K ($_{sat}$ = 5 s).

3.6. STD NMR Experiments to Study the Interaction of Peptides and Glycolipid Micelles

1. NMR assignments: 1D proton spectra as well as 2D TOCSY, COSY, NOESY, and ROESY experiments in 90% H_2O/10% D_2O and 100% D_2O were recorded using the spectrometer's standard pulse sequences (with presaturation for water suppression) (see Note 13) at 25 °C. Complete ^1H assignments were obtained for the peptide. T1 relaxation times were recorded using the standard inversion-recovery setup.
2. STD spectra (Fig. 5) for the peptide were recorded in 100% D_2O and 90% H_2O/10% D_2O. As described above for the ganglioside, an STD difference spectrum was recorded for the peptide in the absence of ganglioside GD2 to confirm that there was no saturation of the peptide at the chosen on-resonance frequency (0 ppm). Reference spectra and a series of STD difference spectra were recorded for different saturation times and peptide–ganglioside concentrations.

3.7. Transferred NOE Experiments (Fig. 6)

1. Control spectrum: 200 ms NOESY of peptide alone. The standard NOESY pulse sequence, with presaturation for suppression of H_2O signal, was used. The relaxation delay was 1.2 s, sweep

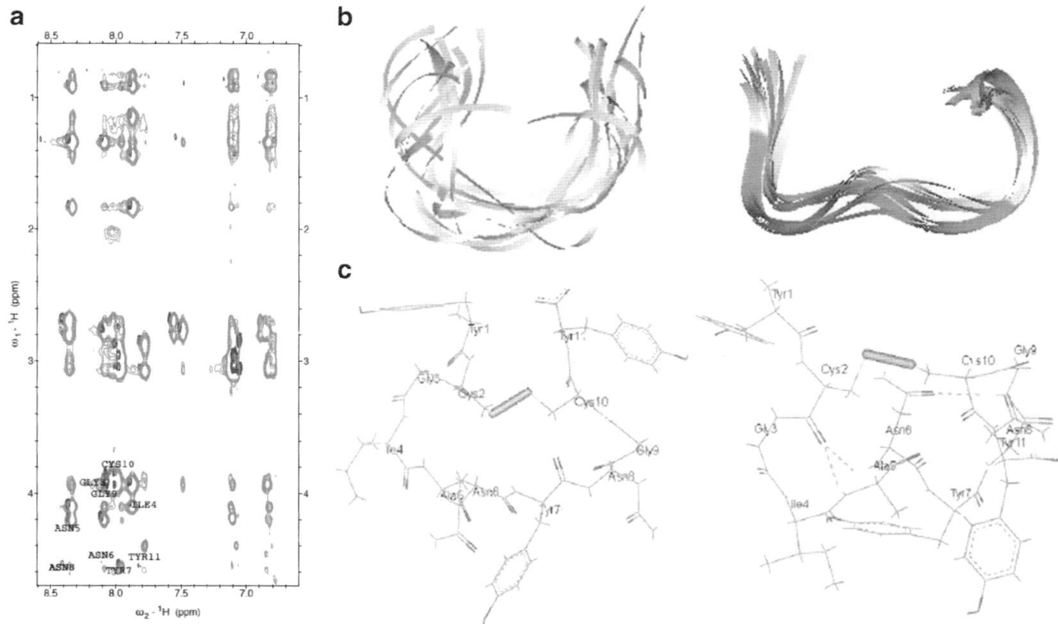

Fig. 6. Ensemble of 12 and 19 lowest-energy structures calculated for the M50B in the free and bound state in water, respectively. (**a**) Overlay of ROESY spectrum (*gray*) of M50B in the free state and trans-NOESY spectrum (*black*) of M50B in the bound state. The mixing times used to record these spectra are 200 and 100 ms, respectively. (**b**) The ensembles of structures are in flat ribbon representation. (**c**) The average structure is in line representation (water molecules are removed). There are three hydrogen bonds between carbonyl group and amide of Cys2 and Ala5, Cys2 and Asn6, and Asn6 and Gly9.

width 16 ppm, and 2,000 points were collected in the direct dimension and 256 complex points in the indirect dimension with a mixing time of 200 ms and 16 transients. If the peptide is unstructured in the absence of binding partner there will be very few cross peaks in this spectrum, and they may be of opposite sign from the diagonal.

2. trNOEs: Similar to the STD experiment, in a weakly binding system, as the ligand is transiently bound to a macromolecule, NOEs between ligand atoms in the bound form develop and this information is maintained in the signal of the free peptide when the peptide dissociates. The same experimental setup is used as for the NOESY of the free ligand. If the peptide becomes structured upon binding macromolecule, there will be many more NOEs visible. The same conditions as described above were used to record an NOESY for the peptide in the presence of ganglioside GD2. The ganglioside resonances are broad enough that there was no need to suppress their signal. A short spin-lock as in the STD experiment can be used to suppress macromolecule signal if it interferes with the 2D spectrum. Additional NOESY spectra with mixing times of 50 and 100 ms were also recorded to assess signal buildup to ensure that correct intensities were used in subsequent structure calculations.

3.8. Structure Calculations of Unbound and Bound Peptide Ligands

1. Cross peaks from NOESY spectra with mixing times of 100 and 200 ms, which lie within the initial buildup of the NOE curve, were measured with SPARKY.
2. Coupling constants (^3JH_N-Hα) measured from 1D proton and 2D COSY spectra were used to constrain the backbone φ angles.
3. The conformations of the peptides are constrained by an intra-chain disulfide between two cysteines.
4. 3D structures of peptide ligands were calculated by the software ARIA 2.2/CNS 1.2 with the network anchoring approach.
5. From the 200 structures calculated for each peptide, the 20 lowest energy ones were chosen and clustered by MMTSB.
6. Structures were visualized using Discovery Studio 1.7, PyMOL, and VMD-Xplor on Mac OS X.

4. Notes

1. Gangliosides, as amphiphilic compounds, spontaneously form micelles (aggregates of high molecular weight) when present in water. Therefore a ganglioside GD2 sugar analog, lacking the lipophilic ceramide functionality, was designed in order to study its conformations and interactions with the antibody in aqueous solution.
2. Intact ganglioside GD2 can be studied through NMR by inserting the whole molecule into DPC or SDS micelles in an appropriate ratio. The sugar head group is solvent exposed and exhibits sharp line-widths in NMR spectra.
3. Prior to recording any NMR spectra, the probe was tuned to the appropriate nuclei, samples were shimmed using automatic gradient shimming, and the deuterium lock was adjusted to be exactly on resonance. 90° proton pulse-widths were measured for each sample, and the transmitter offset for H_2O suppression was optimized.
4. It has been observed that STD intensities can be influenced by the ligand proton's T1 relaxation rate (11). At longer saturation times signals with longer T1s will appear enhanced with respect to signals arising from protons with shorter T1s. Therefore knowledge of the T1 relaxation rates of the individual protons in a ligand is important in order to assess the relative importance of different signals in an STD spectrum. It has been proposed that the rate at which the STD signal builds up at short relaxation times may be a more accurate indication of a binding epitope. In practice it may not always be possible to obtain good signal-to-noise in spectra recorded with very short saturation times.

5. The pulse sequence used for this study was kindly shared by Albin Otter, University of Alberta. Since this time a version of the STD pulse sequence has been included in the VNMR BioPack Pulse sequence library. It is very similar to the one used in this study, with DPFGSE solvent suppression instead of WATERGATE.
6. Different mixing times of NOESY spectra were used to evaluate the linear buildup of NOEs.
7. There are many other docking programs such as AUTODOCK that can be used.
8. The homology structure of Fab fragment of 3F8 was built up based on crystal structure of malaria antigen AMA1 antibody Fab (pdb code 2q8a) (24) and 13G5 antibody Fab (pdb code 2gjz) (25) selected as templates for light chain (84%) and heavy chain (84%) of 3F8, respectively. If there are crystal or NMR structures of the protein of interest available, they can be directly input into the docking programs.
9. Software such as DiscoveryStudio can be used to determine amino acid residues within the binding pocket.
10. *R*-factor can be used as quality control of docking model. The smaller the *R*-factor (<0.5), the more reflective of the true binding mode, providing more accurate information for drug design.
11. The 1D spectrum of ganglioside GD2 showed that it formed micelles at the concentrations used in the STD and trNOE experiments. Line-widths were very broad in comparison to thiophenyl GD2 or the sugar resonances from ganglioside GD2 embedded in DPC micelles. The cmc for ganglioside GD2 has not been reported; however values for related gangliosides range from 3.9×10^{-8} M to 3.4×10^{-9} M (10), well below the lowest concentration of GD2 used in this study.
12. Solubility and pH of peptide solution are very important for NMR experiments. The pH of the final peptide solution in water should be checked, as it is difficult to remove traces of trifluoroacetic acid (TFA) after reverse-phase HPLC (a common step in peptide purification). Presence of TFA may result in lower than expected pH upon dissolution of the peptide. It is important to ensure that the peptide is completely soluble. If needed, the pH or the concentration may be changed to improve solubility.
13. Solvent presaturation was used for 2D spectra; in general watergate should provide equivalent or superior solvent suppression; however on our system we find that due to longer gradient recovery times on the cryogenic probe, presaturation provides better solvent suppression despite lower signal intensity of exchangeable protons.

References

1. Pellecchia M, Bertini I, Cowburn D, et al. Perspectives on NMR in drug discovery: a technique comes of age. Nat Rev Drug Discov. 2008;7:738–45.
2. Meyer B, Peters T. NMR spectroscopy techniques for screening and identifying ligand binding to protein receptors. Angew Chem Int Ed Engl. 2003;42:864–90.
3. Markwick PR, Malliavin T, Nilges M. Structural biology by NMR: structure dynamics and interactions. PLoS Comput Biol. 2008;4:e1000168.
4. Tong W, Gagnon M, Sprules T, et al. Small-molecule ligands of GD2 ganglioside designed from NMR studies exhibit induced-fit binding and bioactivity. Chem Biol. 2010;17:183–94.
5. Modak S, Cheung NK. Disialoganglioside directed immunotherapy of neuroblastoma. Cancer Invest. 2007;25:67–77.
6. Hakomori S-I, Zhang Y. Glycosphingolipid antigens and cancer therapy. Chem Biol. 1997;4:97–104.
7. Birklé S, Zeng G, Gao L, et al. Role of tumor-associated gangliosides in cancer progression. Biochimie. 2003;85:455–63.
8. Jayalakshmi V, Krishna NR. Complete relaxation and conformational exchange matrix (CORCEMA) analysis of intermolecular saturation transfer effects in reversibly forming ligand-receptor complexes. J Magn Reson. 2002;155:106–18.
9. Saragovi HU, Greene MI, Chrusciel RA, Kahn M. Loops and secondary structure mimetics: development and applications in basic science and rational drug design. Biotechnol (N Y). 1992;10:773–8.
10. Sonnino S, Cantù L, Corti M, et al. Aggregative properties of gangliosides in solution. Chem Phys Lipids. 1994;71:21–45.
11. Yan J, Kline AD, Mo H, et al. The effect of relaxation on the epitope mapping by saturation transfer difference NMR. J Magn Reson. 2003;163:270–6.
12. Wang J, Wang W, Kollman PA, Case DA. Automatic atom type and bond type perception in molecular mechanical calculations. J Mol Graph Model. 2006;25:247–60.
13. Case DA, Cheatham 3rd TE, Darden T, et al. The Amber biomolecular simulation programs. J Comput Chem. 2005;26:1668–88.
14. Rieping W, Habeck M, Bardiaux B, et al. ARIA2: Automated NOE assignment and data integration in NMR structure calculation. Bioinformatics. 2007;23:381–2.
15. Feig M, Karanicolas J, Brooks CL. MMTSB Tool Set: enhanced sampling and multiscale modeling methods for applications in structural biology. J Mol Graph Model. 2004;22:377–95.
16. Delaglio F, Grezesiek S, Vuister GW, et al. NMRpipe: a multidimensional spectral processing system based on UNIX pipes. J Biomol NMR. 1995;6:277–93.
17. Houliston RS, Yuki N, Hirama T, et al. Recognition characteristics of monoclonal antibodies that are cross-reactive with gangliosides and lipooligosaccharide from *Campylobacter jejuni* strains associated with Guillain-Barre and Fisher syndromes. Biochemistry. 2007;46: 36–44.
18. Bernatchez S, Szymanski CM, Ishiyama N, et al. A single bifunctional UDP-GlcNAc/Glc 4-Epimerase supports the synthesis of three cell surface glycoconjugates in *Campylobacter jejuni*. J Biol Chem. 2005;280:4792–802.
19. Blixt O, Vasiliu D, Allin K, et al. Chemoenzymatic synthesis of 2-azidoethyl-ganglio-oligosaccharides GD3, GT3, GM2, GD2, GT2, GM1, and GD1a. Carbohydr Res. 2005;340:1963–72.
20. Mayer M, Meyer B. Group epitope mapping by saturation transfer difference NMR to identify segments of a ligand in direct contact with a protein receptor. J Am Chem Soc. 2001;123: 6108–17.
21. Siebert HC, Reuter G, Schauer R, et al. Solution conformations of GM3 ganglioside containing different sialic acid residues as revealed by NOE-based distance mapping molecular mechanics and molecular dynamics calculations. Biochemistry. 1992;31:6962–71.
22. Jakalian A, Jack DB, Bayly CI. Fast efficient generation of high-quality atomic charges AM1-BCC model: II Parameterization and validation. J Comput Chem. 2002;23: 1623–41.
23. Naim M, Bhat S, Rankin KN, et al. Solvated interaction energy (SIE) for scoring protein-ligand binding affinities. 1. Exploring the parameter space. J Chem Inf Model. 2007;47: 122–33.
24. Coley AM, Parisi K, Masciantonio R, et al. The most polymorphic residue on *Plasmodium falciparum* apical membrane antigen 1 determines binding of an invasion-inhibitory antibody. Infect Immun. 2006;74:2628–36.
25. Muller R, Debler EW, Steinmann M, et al. Bifunctional catalysis of proton transfer at an antibody active site. J Am Chem Soc. 2007;129: 460–1.

Chapter 5

A Combinatorial Strategy for the Acquisition of Potent and Specific Protein Tyrosine Phosphatase Inhibitors

Sheng Zhang, Lan Chen, David S. Lawrence, and Zhong-Yin Zhang

Abstract

Protein tyrosine phosphatases (PTPs), a large family of signaling enzymes, regulate many cellular processes, such as proliferation, differentiation, migration, apoptosis, and immune responses. Small molecule inhibitors against PTPs are valuable both as powerful tools to study the functions of target PTPs and as lead compounds for pharmacological development. Here, we describe a novel combinatorial library approach to target simultaneously both the active site pocket and a peripheral secondary binding site for the acquisition of potent and specific PTP inhibitors. Fluorescence tagging during combinatorial library synthesis enables fluorescence polarization-based high-throughput screening of the resulting library, leading to identification of a TC-PTP inhibitor.

Key words: Protein tyrosine phosphatases (PTPs), PTP inhibitor, Peripheral binding site, Fluorescence tagged, Fluorescence polarization, High-throughput screening

1. Introduction

Protein tyrosine phosphatases (PTPs), which catalyze the dephosphorylation of phosphotyrosine-containing proteins, play important roles in a broad range of cellular processes, such as proliferation, differentiation, migration, apoptosis, and immune responses (1, 2). Malfunction of PTP activity is known to be associated with cancer, metabolic syndromes, and autoimmune disorders (3). Small molecule inhibitors of the PTPs not only can be used as research tools to study the function of these enzymes in normal physiology and in diseases, but also can be potential leads for therapeutic development. However, achieving specificity for PTP inhibition is not trivial, particularly when the target is a member of a large protein family. Indeed, the common architecture of PTP active sites impedes the development of selective PTP inhibitors. Fortunately, PTP substrate specificity studies have shown that pTyr

Yi Zheng (ed.), *Rational Drug Design: Methods and Protocols*, Methods in Molecular Biology, vol. 928,
DOI 10.1007/978-1-62703-008-3_5,

alone is not sufficient for high-affinity binding and residues flanking pTyr are important for PTP substrate recognition (4). These results suggest that unique PTP sub-pockets that border the active site (i.e., the pTyr-binding site) may be targeted to enhance inhibitor potency and selectivity. In principle, active site-directed, potent and selective PTP inhibitors can be devised by tethering a nonhydrolyzable pTyr mimetic to appropriately functionalized moieties to engage both the active site and unique nearby sub-pockets (Fig. 1) (5). In this bidentate inhibitor approach, the pTyr mimetic may interact in the desired inhibitory fashion with all PTPs; the molecular scaffolds attached to it should render the inhibitors PTP isozyme-selective. Moreover, the peripheral site binder will also enhance inhibitor potency due to additivity of free energy of binding. Utilizing this strategy we have obtained several PTP inhibitors with desired potency as well as specificity in a relatively efficient way in terms of both time and cost (6–10). Here, we will describe the design of a bidentate PTP inhibitor library and a fluorescence-tagged combinatorial synthesis and fluorescence polarization-based high-throughput screening, which we used to obtain a potent and selective TC-PTP inhibitor (8).

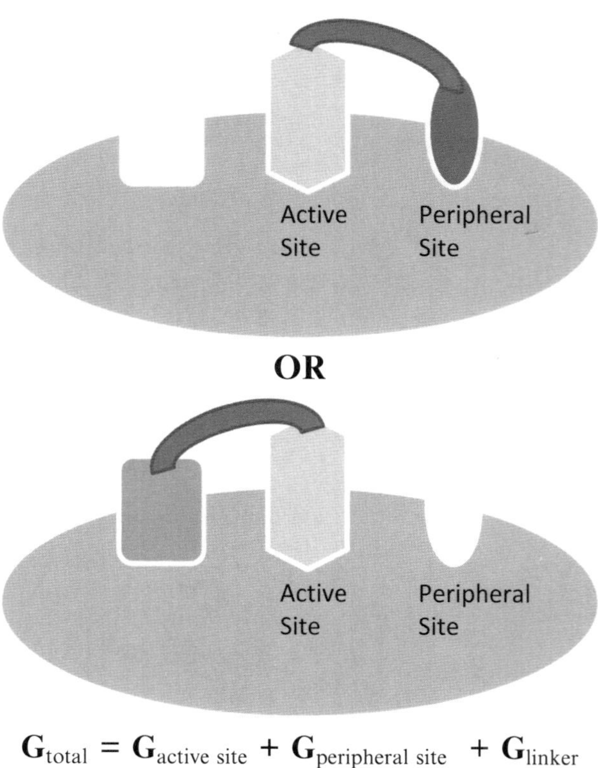

Fig. 1. "Peripheral Site" strategy for acquiring potent and selective PTP inhibitors.

1.1. Bidentate Inhibitor Library Design

The basic structure of the bidentate PTP inhibitors is shown in Fig. 2, with phosphonodifluoromethyl phenylalanine (F_2Pmp) targeting the PTP catalytic pocket. As shown in this figure, the bidentate inhibitor contains three components: (a) F_2Pmp as the active site-directed motif, (b) an R group as the peripheral site-binding motif, and (c) the linker that connects the F_2Pmp and the R group. Other bidentate inhibitor libraries can be designed similarly, with different active site targeting motifs and/or different linkers. The library of potential bidentate PTP inhibitors can be prepared by condensation reaction of the common core (compound 1 in Fig. 3)

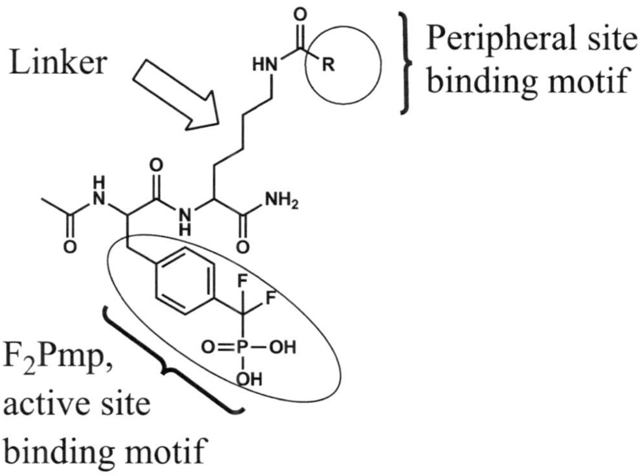

Fig. 2. Structure of the bidentate PTP inhibitor. The bidentate PTP inhibitor contains F2Pmp as the active site-directed motif (shown in *oval*), an R group as the peripheral site-binding motif (shown in *circle*), and a linker.

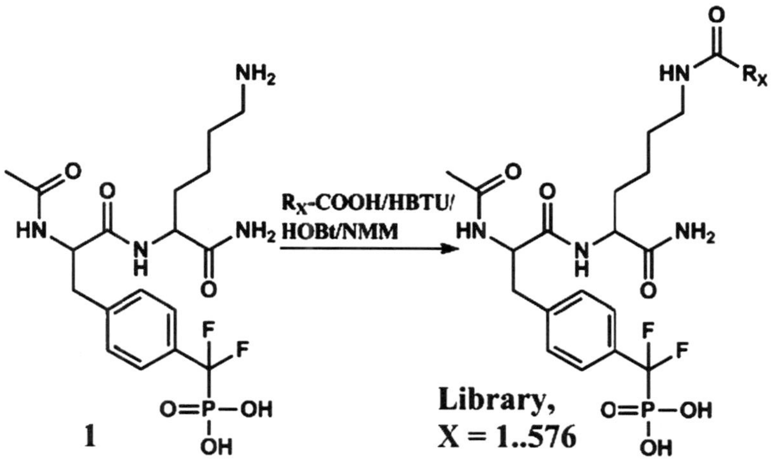

Fig. 3. Synthesize the bidentate ligand library by condensing core **1** with different carboxylic acids.

with different carboxylic acids. In addition to condensation reactions, the library can be constructed using a variety of assembly modalities (e.g., "Click Chemistry") (7, 9, 10) either in solution or on solid support (6). Once prepared, the library can then be screened against the target PTP in order to identify lead inhibitory species. The latter typically need to be resynthesized on a larger scale for detailed studies to confirm inhibitory activity and evaluate intracellular behavior.

The fluorophore-tagged compound library compounds are designed in the same way as described above, except that a fluorescein tag is attached to each library member at a position separated from critical inhibitory functionality (Fig. 4). The fluorophore allows the binding affinity of the library compounds for the target PTP to be measured using a fluorescence polarization (FP)-based displacement assay (11). Since nonfluorescent impurities do not affect the assay, the library can be screened as is without the need for further purification, an otherwise time-consuming and laborious process. In addition, the FP value is an intensive property (as versus to extensive property), which means it is independent of the concentration of fluorophore-tagged compound. As a result, any variation in the concentration of the compound will not bias the screening result. Finally, we note that the fluorophore-tagged library can be prepared in a fashion analogous to that of the corresponding non-tagged library (Fig. 5).

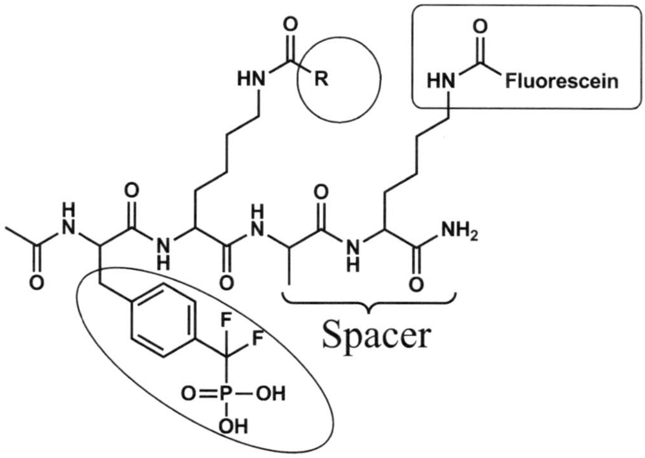

Fig. 4. Structure of bidentate PTP inhibitor library with a fluorescein tag. Each library member contains F2Pmp as the active site-directed motif (shown in *oval*), an R group as the peripheral site-binding motif (shown in *circle*), and a fluorescein tag (shown in *rectangle*).

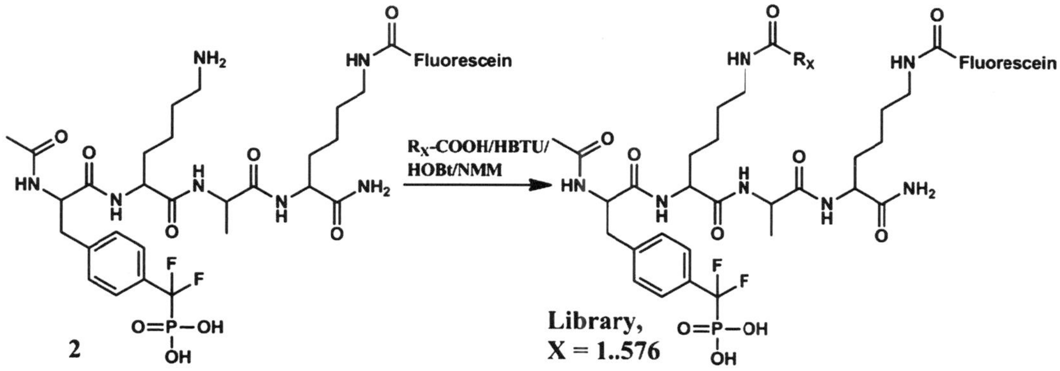

Fig. 5. Synthesize the fluorescein-tagged library by condensing core **2** with different carboxylic acids.

2. Materials

Prepare all the aqueous solutions with Milli-Q water. Prepare all the organic solutions with certified ACS grade solvents. All the organic solvents and reagents should be handled in a fume hood. Avoid any contact or inhalation.

2.1. DMF Solution of the Coupling Reagents

1. HBTU solution (Advanced ChemTech): 35 mM HBTU in DMF (Fisher Scientific). Weigh 1.32 g HBTU and transfer to a dry Erlenmeyer flask. Add 100 mL DMF to the flask, and gently shake the flask until all the HBTU dissolves. Transfer the HBTU solution into a glass bottle. Use a cap with TFE liner to cap the bottle tightly, and seal it with Parafilm. Store the HBTU solution at –20°C (see Note 1).
2. HOBt solution (Advanced ChemTech): 50 mM HOBt in DMF. Weigh 0.76 g HOBt monohydrate and transfer to a dry Erlenmeyer flask. Add 100 mL DMF, and gently shake the flask until all the HOBt dissolves. Transfer the HOBt solution into a glass bottle and cap it tightly. Store the HOBt solution at –20°C.
3. NMM solution (Fisher Scientific): 200 mM NMM in DMF. Weigh 2.02 g NMM and transfer to a dry Erlenmeyer flask. Add 100 mL DMF, and gently shake the flask to mix the NMM with DMF. Transfer the NMM solution into a glass bottle and cap it tightly. Store the NMM solution at –20°C.
4. Cyclohexylamine solution (Aldrich): 87 mM in DMF. Weigh 0.86 g cyclohexylamine and transfer to a dry Erlenmeyer flask. Add 100 mL DMF and gently shake the flask to mix the cyclohexylamine with DMF. Transfer the cyclohexylamine solution into a glass bottle and cap it tightly. Store the cyclohexylamine solution at –20°C.

5. The library core **2**: Synthesize compound **2** using standard solid-phase peptide synthesis (SPPS) protocol (8). Following the synthesis, purify it by HPLC. Make a 2 mM DMF solution of the core **2** and store the solution at –20°C.
6. Carboxylic acid library: Make a 40 mM DMF solution for each acid, and transfer the acid solution into a 96-well polypropylene DeepWell plate (NUNC). Seal the plates with aluminum sealing tape (NUNC) and store the plates at 4°C (see Note 2).
7. DMG buffer: 50 mM DMG, pH 7.0. Weigh 4.01 g 3,3-dimethyl glutaric acid, 0.52 g sodium chloride, and 0.186 g ethylenediaminetetraacetic acid (EDTA) and transfer to a 1-L graduated cylinder. Add 400 mL Milli-Q water into the cylinder. Gently shake the graduated cylinder until all the solid dissolve. Adjust pH to 7.0 using NaOH. Make up the volume to 500 mL with Milli-Q water. Store the DMG buffer at room temperature.
8. The target PTP protein with GST tag can be over-expressed in *Escherichia coli* and purified using GST-tag purification resins. Add 30% glycerol to the purified protein and store the protein at –20°C.

3. Methods

Carry out all procedures on a Freedom EVO workstation (Tecan) with a 96-channel MCA tip block using disposable tips. The layout of the workstation with the plates and reagent troughs is shown in Fig. 6. The complete EVO program is listed in Note 3.

3.1. Combinatorial Synthesis of Fluorescence-Tagged Library

1. Put four polypropylene troughs on the Freedom Evo workstation. Label them as "HBTU," "HOBt," "NMM," and "Cyclohexylamine." Fill them with 30 mL HBTU, HOBt, NMM, and cyclohexylamine solution accordingly (see Note 4).

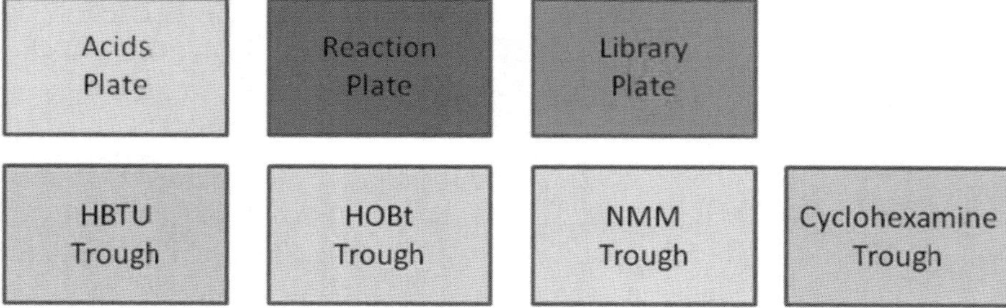

Fig. 6. Suggested layout for the troughs and plates for the library synthesis.

2. Take a 96-well polypropylene plate, and label it as "Library." Manually add 10 μL of library core **2** solution to each well. Put the "Library" plate on the Freedom Evo workstation.
3. Put another empty 96-well polypropylene plate on the Freedom Evo workstation, and label it as "Reaction."
4. Put the 96-well DeepWell plate of carboxylic acids on the Freedom Evo workstation.
5. Load the 96-channel MCA tip block with new disposable tips. Transfer 10 μL of carboxylic acid solution from the 96-well DeepWell plate of the carboxylic acid library to the empty "Reaction" plate. Dispose the tips.
6. Load the tip block with new tips. Transfer 10 μL of HBTU solution from the "HBTU" trough to the "Reaction" plate. Mix the solution by pipetting up and down three times. Dispose the tips. Start a timer with 5 min countdown time.
7. Load the tip block with new tips. Transfer 10 μL of HOBt solution from the "HOBt" trough to the "Reaction" plate. Mix the solution by pipetting up and down three times. Dispose the tips.
8. After the 5 min countdown time is over, load the tip block with new tips. Transfer 10 μL of NMM solution from the "NMM" trough to the "Reaction" plate. Mix the solution by pipetting up and down three times. Transfer all 40 μL solution from the "Reaction" plate to the "Library" plate. Mix the solution by pipetting up and down three times. Dispose the tips.
9. After 1 h, load the tip block with new tips. Transfer 10 μL of cyclohexylamine solution from the "Cyclohexylamine" trough to the "Library" plate. Mix the solution by pipetting up and down three times. Dispose the tips.
10. After 30 min, manually add 190 μL DMSO to each well of the "Library" plate using a multichannel pipette. Mix the solution by pipetting up and down three times. Dispose the tips. Seal the "Library" plate with aluminum sealing tape and store the plate at −20°C. The final concentration of the fluorescence-tagged library compound is about 80 μM.
11. Repeat steps 2–10 for all the plates of the carboxylic acid library.

3.2. Fluorescence Polarization-Based High-Throughput Screening

1. The fluorescence-tagged library compound from the DMSO solution should be diluted with DMG buffer, until the concentration of the library compound is about 75 nM. Store the diluted library (75 nM) in a 96-well plate.
2. Dilute the PTP protein stock to 2 μM with DMG buffer. About 20 mL of the 2 μM PTP solution is enough for four 96-well plates of library compounds (384 compounds).

3. Take a black 384-well polystyrene plate, and label it as "assay" plate. Add 50 μL 2 μM PTP solution to each well of the "assay" plate using Multidrop. Use Freedom EVO workstation to transfer 2 μL library compounds (75 nM stock) to the "assay" plate. Mix the solution by pipetting up and down three times (see Note 5).
4. Use EnVision to measure the fluorescence anisotropy value (A_1) of each well of the assay plate (see Note 6). The anisotropy value of each library compound can be exported to an Excel file.
5. Prepare another 2 μM protein solution with 0.5 mM Fmoc-F_2Pmp-OH, a known active center binding PTP inhibitor. Repeat steps 3 and 4. The result fluorescence anisotropy value (A_2) of each compound can be exported to another Excel file.
6. For each compound, check the A_1 value: If the A_1 is larger than 50 mA, calculate the displacement percentage using the equation: displacement percentage (%) = $(A_1 - A_2)/(A_1 - 30) \times 100\%$; if the A_1 is smaller than 50 mA, set the displacement percentage (%) as 100% (see Note 7). Sort the compound based on the displacement percentage. The compounds with the smallest displacement percentage value are the hits from the screening.
7. The hit compounds need to be resynthesized without the fluorescence tag and their inhibitory activity needs to be confirmed in the activity-based enzymatic assay.

4. Notes

1. All glassware should be dried in an oven (>100°C) overnight and cooled to room temperature in a desiccator before use. The HBTU solution can be stored for months if it is free of water and sealed properly with Parafilm.
2. The plate should be sealed carefully with the sealing tape to ensure that every well is tightly sealed, so that the acid solution will not absorb moisture. For better protection, the sealed plate can be wrapped with three layers of plastic food wrap.
3. EVOware program for the automatic synthesis of library.

```
Comment Begin of the library synthesis
1
Comment Transfer acid to the Reaction Plate
2
Get DiTis Grid 1; Site: 1 (200ul RaininTips
TecanBox)
Fetch 8 rows and 12 columns
3
```

```
Aspirate 10 µl SZ-DMF
"Acid Plate" (Col. 1, Rows 1-8)
4
Dispense 10 µl SZ-DMF
"Reaction Plate" (Col. 1, Rows 1-8)
5
Drop DiTis Grid 1; Site: 1 (200ul RaininTips
TecanBox)
6
Comment Add HBTU to the Reaction Plate
7
Get DiTis Grid 1; Site: 2 (200ul RaininTips
TecanBox)
Fetch 8 rows and 12 columns
8
Aspirate 10 µl SZ-DMF
"HBTU Trough" (Col. 1, Rows 1-8)
9
Dispense 10 µl SZ-DMF
"Reaction Plate" (Col. 1, Rows 1-8)
10
Begin Loop 3 times "Mixing"
11
Aspirate 15 µl SZ-DMF
"Reaction Plate" (Col. 1, Rows 1-8)
12
Dispense 15 µl SZ-DMF
"Reaction Plate" (Col. 1, Rows 1-8)
13
End Loop "Mixing"
14
Drop DiTis Grid 1; Site: 2 (200ul RaininTips
TecanBox)
15
Start Timer 1
16
Comment Add HOBt to the Reaction Plate
17
Get DiTis Grid 1; Site: 3 (200ul RaininTips
TecanBox)
Fetch 8 rows and 12 columns
18
Aspirate 10 µl SZ-DMF
"HOBt Trough" (Col. 1, Rows 1-8)
19
Dispense 10 µl SZ-DMF
"Reaction Plate" (Col. 1, Rows 1-8)
20
Begin Loop 3 times "Mixing"
```

```
21
Aspirate 25 µl SZ-DMF
"Reaction Plate" (Col. 1, Rows 1-8)
22
Dispense 25 µl SZ-DMF
"Reaction Plate" (Col. 1, Rows 1-8)
23
End Loop "Mixing"
24
Drop DiTis Grid 1; Site: 3 (200ul RaininTips
TecanBox)
25
Wait for Timer Timer 1: 300 sec
26
Comment Add NMM to the Reaction Plate
27
Get DiTis Grid 7; Site: 1 (200ul RaininTips
TecanBox)
Fetch 8 rows and 12 columns
28
Aspirate 10 µl SZ-DMF
"NMM Trough" (Col. 1, Rows 1-8)
29
Dispense 10 µl SZ-DMF
"Reaction Plate" (Col. 1, Rows 1-8)
30
Begin Loop 3 times "Mixing"
31
Aspirate 35 µl SZ-DMF
"Reaction Plate" (Col. 1, Rows 1-8)
32
Dispense 35 µl SZ-DMF
"Reaction Plate" (Col. 1, Rows 1-8)
33
End Loop "Mixing"
34
Comment Add reaction solution to the Library
Plate
35
Aspirate 40 µl SZ-DMF
"Reaction Plate" (Col. 1, Rows 1-8)
36
Dispense 40 µl SZ-DMF
"Library Plate" (Col. 1, Rows 1-8)
37
Begin Loop 3 times "Mixing"
38
Aspirate 40 µl SZ-DMF
```

```
"Library Plate" (Col. 1, Rows 1-8)
39
Dispense 40 µl SZ-DMF
"Library Plate" (Col. 1, Rows 1-8)
40
End Loop "Mixing"
41
Drop DiTis Grid 7; Site: 1 (200ul RaininTips
TecanBox)
42
Start Timer 2
43
Wait for Timer Timer 2: 3600 sec
44
Comment Add cyclohexylamine to the Library
Plate
45
Get DiTis Grid 7; Site: 2 (200ul RaininTips
TecanBox)
Fetch 8 rows and 12 columns
46
Aspirate 10 µl SZ-DMF
"Cyclohexylamine Trough" (Col. 1, Rows 1-8)
47
Dispense 10 µl SZ-DMF
"Library Plate" (Col. 1, Rows 1-8)
48
Begin Loop 3 times "Mixing"
49
Aspirate 50 µl SZ-DMF
"Library Plate" (Col. 1, Rows 1-8)
50
Dispense 50 µl SZ-DMF
"Library Plate" (Col. 1, Rows 1-8)
51
End Loop "Mixing"
52
Drop DiTis Grid 7; Site: 2 (200ul RaininTips
TecanBox)
53
Comment End of the library synthesis
54
```

4. If the DMF solution is stored in freezer, allow it to warm up to room temperature before pouring into trough, because the cold solution will absorb moisture while warming up.
5. The library compounds are in 96-well plates and will be reformatted to 384-well assay plates for screening. The relationship

of the two types of plates is as follows: for 96-well plate no. 1, the well "A1" goes to "A1" of 384-well pate; for 96-well plate no. 2, the well "A1" goes to "A2" of 384-well pate; for 96-well plate no. 3, the well "A1" goes to "B1" of 384-well pate; and for 96-well plate no. 4, the well "A1" goes to "B2" of 384-well pate. After screening, the result (associated with 384-well format) needs to be assigned to each of the library compound (associated with 96-well format) according to this reformatting relationship.

6. Most plate readers can give both fluorescence polarization (*P*) value and anisotropy (*A*) value as result. For this screening, the fluorescence anisotropy value should be used instead of fluorescence polarization value, because only the fluorescence anisotropy, but not fluorescence polarization, holds the linear superposition principle with each fluorophore component in solution (12). Therefore, the displacement percentage can only be calculated using the anisotropy values.
7. Occasionally, there may be some library compounds that do not bind to the target protein. These compounds will give small A_1 and A_2 value (both close to 30 mA). So we set their displacement percentage as 100% to prevent these compounds from being selected as hits.

Acknowledgement

This work was supported by NIH grant RO1 CA126937, RO1 CA152194 RO1 CA69202, and RO1 CA79954.

References

1. Alonso A, Sasin J, Bottini N, Friedberg I, Osterman A, Godzik A, Hunter T, Dixon J, Mustelin T (2004) Protein tyrosine phosphatases in the human genome. Cell 117:699–711
2. Tonks NK (2006) Protein tyrosine phosphatases: from genes, to function, to disease. Nat Rev Mol Cell Biol 7:833–846
3. Zhang ZY (2001) Protein tyrosine phosphatases: prospects for therapeutics. Curr Opin Chem Biol 5:416–423
4. Zhang ZY (2002) Protein tyrosine phosphatases: structure and function, substrate specificity, and inhibitor development. Annu Rev Pharmacol Toxicol 42:209–234
5. Zhang ZY (2003) Chemical and mechanistic approaches to the study of protein tyrosine phosphatases. Acc Chem Res 36:385–392
6. Shen K, Keng YF, Wu L, Guo XL, Lawrence DS, Zhang ZY (2001) Acquisition of a specific and potent PTP1B inhibitor from a novel combinatorial library and screening procedure. J Biol Chem 276:47311–47319
7. Yu X, Sun JP, He Y, Guo X, Liu S, Zhou B, Hudmon A, Zhang ZY (2007) Structure, inhibitor, and regulatory mechanism of Lyp, a lymphoid-specific tyrosine phosphatase implicated in autoimmune diseases. Proc Natl Acad Sci USA 104:19767–19772
8. Zhang S, Chen L, Luo Y, Gunawan A, Lawrence DS, Zhang ZY (2009) Acquisition of a potent and selective TC-PTP inhibitor via a stepwise fluorophore-tagged combinatorial synthesis and screening strategy. J Am Chem Soc 131:13072–13079

9. Zhang X, He Y, Liu S, Yu Z, Jiang ZX, Yang Z, Dong Y, Nabinger SC, Wu L, Gunawan AM, Wang L, Chan RJ, Zhang ZY (2010) Salicylic acid based small molecule inhibitor for the oncogenic Src homology-2 domain containing protein tyrosine phosphatase-2 (SHP2). J Med Chem 53:2482–2493
10. Zhou B, He Y, Zhang X, Xu J, Luo Y, Wang Y, Franzblau SG, Yang Z, Chan RJ, Liu Y, Zheng J, Zhang ZY (2010) Targeting mycobacterium protein tyrosine phosphatase B for antituberculosis agents. Proc Natl Acad Sci USA 107:4573–4578
11. Zhang S, Chen L, Kumar S, Wu L, Lawrence DS, Zhang ZY (2007) An affinity-based fluorescence polarization assay for protein tyrosine phosphatases. Methods 42: 261–267
12. Roehrl MH, Wang JY, Wagner G (2004) A general framework for development and data analysis of competitive high-throughput screens for small-molecule inhibitors of protein-protein interactions by fluorescence polarization. Biochemistry 43:16056–16066

Chapter 6

Identification of Allosteric Inhibitors of p21-Activated Kinase

Julien Viaud and Jeffrey R. Peterson

Abstract

Protein kinases are among the most important drug targets; however the structural conservation of the ATP-binding pocket of kinases can lead to promiscuous inhibition of additional unintended kinase targets. Allosteric inhibitors that target less conserved regions of protein kinases represent an alternative approach that may provide more selective kinase inhibition. In this report, protocols are provided for the screening and identification of Pak1 inhibitors acting via an allosteric mechanism.

Key words: Kinase inhibitor, Protein kinase, Pak kinase, Autoinhibition, Allosteric inhibition, Cdc42, Rho GTPases, GTP-binding protein

1. Introduction

Target specificity is a major challenge for kinase inhibitors due to the high degree of conservation in the ATP-binding pocket (1). The presence of more unique regulatory domains in some kinases could represent alternative inhibitor targets to achieve greater kinase selectivity. Kinase inhibitor screening programs frequently use constitutively active kinases and may utilize kinase domains in isolation, therefore biasing such screens toward ATP-competitive inhibitors and limiting the discovery of compounds acting via novel allosteric mechanisms. Recently, we described the identification of an allosteric inhibitor targeting Pak1 (p21-activated kinase 1) (2, 3). Pak1 is a cytoplasmic serine/threonine kinase whose hyperactivity has been linked to tumorigenesis (4–7). In its inactive form, Pak1 adopts a homodimeric state in which the N-terminal regulatory domain of one monomer binds and inhibits the catalytic activity of the C-terminal kinase domain of the other monomer and vice versa. Binding of the small G proteins Rac1 or Cdc42 to the regulatory domain leads to a conformational change and kinase activation (8, 9).

Yi Zheng (ed.), *Rational Drug Design: Methods and Protocols*, Methods in Molecular Biology, vol. 928, DOI 10.1007/978-1-62703-008-3_6, © Springer Science+Business Media New York 2012

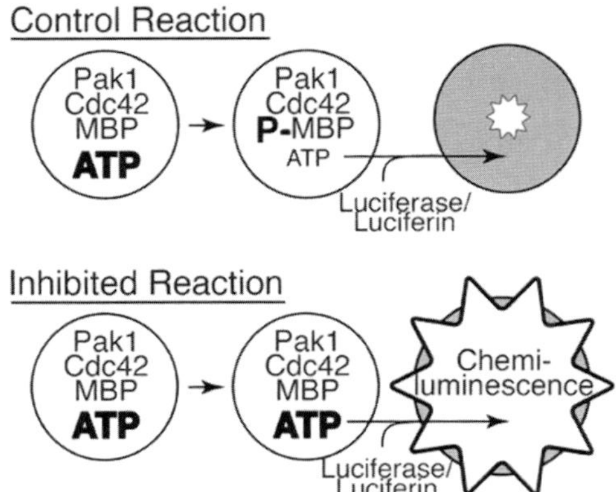

Fig. 1. Pak1 screening assay scheme. Pak1 is incubated with its activator, GTPγS-charged Cdc42, and a substrate, myelin basic protein (MBP), in the presence of 10 μM ATP. Control kinase reactions (*top*) hydrolyze a substantial fraction of the starting ATP resulting in low final ATP concentrations, whereas inhibited reactions (*bottom*) do not. Residual ATP is enzymatically converted to chemiluminescence proportional to residual ATP (reproduced from ref. 2 with permission from Cell Press).

More recently, phosphoinositides have also been shown to play an essential role in Pak1 activation in vivo (10).

To find Pak1 inhibitors that exploit this allosteric regulatory mechanism, we developed a high-throughput assay that recapitulates Cdc42-dependent Pak1 activation, monitoring ATP hydrolysis as an indicator of catalytic activity (see Fig. 1). In this chapter, we report the detailed methodology used for the screening. This assay was used to screen 33,000 structurally diverse small molecules in duplicate. Full-length Pak1 in its inactive form was first incubated with individual chemical compounds and then active Cdc42 and myelin basic protein (MBP) as substrate were added in the presence of 10 μM ATP. After incubation to allow Pak1 activation and MBP phosphorylation, residual nonhydrolyzed ATP was quantified using Kinase-Glo to monitor Pak1 activity. This initial screen was designed to capture both allosteric and ATP-competitive inhibitors of Pak1. 342 active compounds were identified (1% hit rate). Active compounds were then retested in a secondary screen in the presence of high concentrations of ATP to select against ATP-competitive inhibitors. Hit compounds from the primary screen were incubated with Pak1, Cdc42, and MBP in the presence of [γ-^{32}P]ATP and 1 mM unlabeled ATP to directly monitor Pak1 kinase activity toward the substrate MBP. Of the initial 342 hits identified in the primary screen, 32 compounds continued to exhibit robust inhibition at 1 mM ATP and represent candidate allosteric *i*nhibitors of *P*ak1 *a*ctivation (IPAs).

The methodology described here allows the screening of ~2,000 compounds in duplicate per day. Screening in duplicate is highly recommended to reduce the identification of false positives and to assess the reproducibility of active compounds.

2. Materials

Prepare all solutions using ultrapure water (resistivity of 18 MΩ cm at 25°C) and analytical grade reagents. Prepare and store all reagents at room temperature (unless indicated otherwise). The materials described here allow the screening of 12 assay plates (1 day worth, six compound plates in duplicate).

2.1. Recombinant Protein Production Components

Protein expression via recombinant baculovirus infection of insect cells offers a protein production system frequently used for eukaryotic proteins, like Pak1, that are not adequately expressed in a functional form in bacteria. In the case of Pak1, insect cell-expressed Pak1 is produced in its inactive, autoinhibited form (11).

2.1.1. HIS-Pak1 Production in Sf9 Cells

1. Heating, rotary shaker incubator.
2. Hemocytometer for counting cell density.
3. Sf9 culture media containing recombinant baculovirus encoding HIS-Pak1 expression.
4. Sf9 cells (for protein production, 500 mL culture volume at $1–2 \times 10^6$ cells/mL should be sufficient to yield 5 mg of protein).
5. Invitrogen Gibco SF900II insect cell growth media supplemented with L-glutamine at 2 mM, sodium benzylpenicillin at 30 μg/mL, and streptomycin sulfate at 50 μg/mL (see Note 1).
6. Sterile shaker culture flasks. Sf9 cells can be grown routinely in medium occupying up to 1/4 of the flask volume, e.g., 500 mL culture in 2 L flasks, to allow adequate aeration.
7. 500 mL PBS (Phosphate-Buffered Saline).
8. Conical 500 mL tubes for Sf9 sedimentation.

2.1.2. HIS-Pak1 Purification

1. Lysis Buffer: 50 mM Sodium Phosphate, 0.5 M NaCl, 5 mM imidazole, 10 μg/mL each of protease inhibitors chymostatin/leupeptin/pepstatin (see Note 2), 1 mM PMSF (see Note 3), adjust pH to 8.0. Prepare 16 mL per liter culture.
2. Branson tip sonifier.
3. Nickel-NTA beads: 2.5 mL volume per liter culture.
4. Sodium Azide.
5. Washing Buffer: 50 mM Sodium Phosphate, 0.5 M NaCl, 20 mM imidazole (adjust pH to 8.0 carefully because imidazole is acidic). Prepare 150 mL.
6. Elution Buffer: 50 mM Sodium Phosphate, 0.5 M NaCl, 250 mM imidazole (adjust pH to 8.0 carefully because imidazole is acidic). Prepare 20 mL.
7. Pak1 Storage Buffer: 50 mM Tris pH 7.5, 100 mM NaCl, 1 mM DTT. Prepare 2 L.

2.1.3. Expression and Purification of Cdc42

1. BL21 *E. coli* expressing GST-Cdc42 (in our case, Cdc42 is cloned into pGEX-4T).
2. Luria Bertani Broth (LB): Prepare a flask containing 50 mL LB for the overnight preculture and a flask containing 950 mL LB for protein production. The flasks containing the LB should be autoclaved and cooled to room temperature prior to inoculation.
3. Antibiotic stock solution (in our case, the pGEX-4T contains an ampicillin resistance gene).
4. Lysis Buffer: PBS containing 1 mM PMSF, 1 mM $MgCl_2$, 25 μM GDP, 0.5 mg/lysozyme. Prepare 20 mL.
5. Glutathione agarose beads.
6. Washing Buffer: PBS containing 1 mM $MgCl_2$, 25 μM GDP. Prepare 30 mL.
7. Thrombin.
8. Benzamidine sepharose beads.

2.1.4. GTPγS Charging of Cdc42

1. 10 mM GTPγS stock solution in water.
2. 0.4 M EDTA stock solution in water.
3. 1 M $MgCl_2$ stock solution.

2.2. Screen Components

1. HIS-Pak1 (see Subheading 2.1.1 for the materials and Subheadings 3.1 and 3.2 for the methods).
2. Cdc42-GTPγS (see Subheading 2.1.3 for the materials and Subheadings 3.3 and 3.4 for the methods).
3. MBP: Prepare 10 mL of a stock solution of 2 mg/mL. For economical preparation of larger amounts, MBP can also be purified from bovine brain acetone powder; see Note 4.
4. ATP: Resuspend the ATP in water. Prepare 2 mL of a stock solution of 500 μM.
5. 1× Phospho Buffer: 50 mM Hepes pH 7.5, 25 mM NaCl, 1.25 mM $MgCl_2$, 1.25 mM $MnCl_2$. Prepare 1 L.
6. Assay plates are 384-well, white, flat bottom, low volume, tissue culture treated, sterile, polystyrene.
7. Desiccator.
8. Compound plates: The compound library we used consisted of 33,000 diverse compounds obtained from ChemDiv and the Challenge Set, Natural Products Set, and Structural Diversity Set provided by the Developmental Therapeutics Program (DTP) of the National Cancer Institute. Compounds in 384-well formats are stored as 5 mM DMSO stock solutions at −80°C.
9. Staurosporine.

10. Kinase Glo®.
11. MicroFill 96-/384-Well Reagent Dispenser.
12. Disposable 384-pin polypropylene pin replicators.
13. Cary Eclipse fluorescence spectrophotometer/luminometer equipped with a microplate carrier.

2.3. ^{32}P-ATP Kinase Assay (Secondary Screen)

1. 1.5 mL locking-lid microcentrifuge tubes.
2. Individual compounds dissolved in DMSO.
3. HIS-Pak1 (see Subheading 2.1.1 for the materials and Subheadings 3.1 and 3.2 for the methods).
4. Cdc42-GTPγS (see Subheading 2.1.3 for the materials and Subheadings 3.3 and 3.4 for the methods).
5. MBP (from Subheading 2.2, item 3).
6. [γ-^{32}P]ATP.
7. 5× Kinase Assay Buffer: 250 mM Hepes pH 7.5, 125 mM NaCl, 6.25 mM $MgCl_2$, 6.25 mM $MnCl_2$.
8. SDS-PAGE Sample Buffer.
9. 95 °C heating block.
10. 15% Polyacrylamide gels for SDS-PAGE.
11. Standard solutions/equipment for SDS-PAGE electrophoresis.
12. Gel dryer.
13. PhosphorImager.

3. Methods

In this section, we describe procedures for the production and purification of HIS-Pak1 and Cdc42-GTPγS (respectively, Subheadings 3.1–3.2 and 3.3–3.4) before describing the preparation of the different solutions and the screening procedure (Subheadings 3.5–3.7).

Overview of the methodology: Recombinant, HIS-tagged Pak1 is incubated with individual compounds followed by addition of recombinant Cdc42-GTPγS and MBP as substrate in the presence of 10 μM ATP. Following incubation to allow Pak activation and MBP phosphorylation, nonhydrolyzed ATP is quantified using Kinase-Glo (12). To define the conditions of the assay, careful titrations of both Pak1 and Cdc42 are required to establish that (a) adequate Pak1 is provided to substantially deplete ATP during the course of the incubation (Fig. 2a) and (b) sufficient Cdc42 is included to activate Pak1 by ~90% (Fig. 2b). By utilizing a sub-saturating Cdc42 concentration (~90% activation) the assay is most

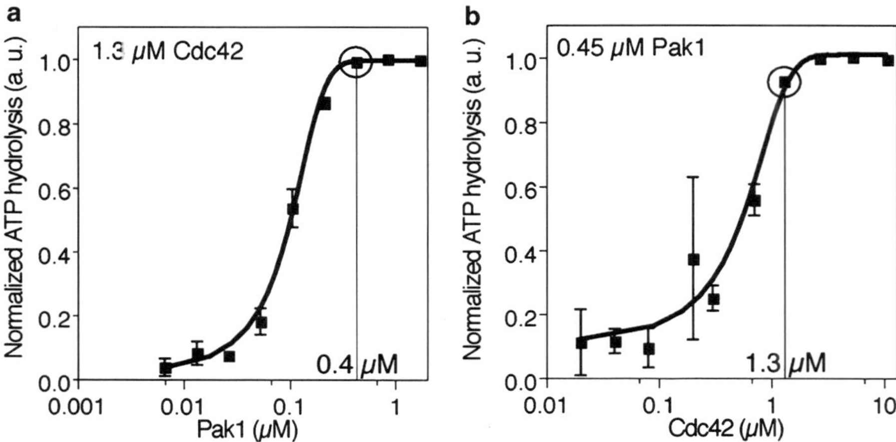

Fig. 2. Titration of Pak1 and Cdc42 concentrations in the screening assay. (**a**) Cdc42 was incubated with MBP, ATP, and the indicated concentrations of Pak1. Residual ATP levels were measured using Kinase Glo (Promega). Results are expressed as ATP hydrolyzed in arbitrary units (normalized to the maximum ATP hydrolyzed) as a function of Pak1 concentration. Data points and error bars show mean and standard error of triplicate wells. Circled point indicates Pak1 concentration used in the screen. (**b**) ATP hydrolysis by Pak1 is presented as a function of Cdc42 concentration in the reaction. Circled point indicates Cdc42 concentration used in the screen (reproduced from ref. 2 with permission from Cell Press).

sensitive to compounds that compete with activation by Cdc42. In addition, these titrations demonstrate that ATP hydrolysis strictly depends on both Pak1 and Cdc42 and confirms that the recombinant Pak1 is autoinhibited yet can be activated by Cdc42, as expected. Since this screening methodology does not differentiate between allosteric and ATP-competitive inhibitors, compounds active in the primary screening assay are retested in a conventional [^{32}P]ATP kinase assay in the presence of 1 mM ATP to select against ATP-competitive compounds (Subheading 3.7).

3.1. Infection of Sf9 Cells for Production of HIS-Pak1

All manipulations should be conducted in a sterile cell culture cabinet using sterile technique. A heated shaking incubator, dedicated to insect cells culture, should be used for Sf9 cell culture to reduce the risk of bacterial contamination. A solution of 10% bleach should be kept handy to decontaminate any spills of live baculovirus.

1. Grow Sf9 cells to a density of 1–2 × 10^6 cells/mL (500 mL culture/2 L flask for large-scale expression) in SF900II medium at 27°C with shaking at 125 rpm.
2. Infect cells by the addition of 1:10 volume of baculovirus stock (e.g.: for 500 mL culture, add 50 mL of baculovirus). Infection with a large-volume fraction of high-titer baculovirus stock ensures simultaneous infection of most cells in the culture.
3. Incubate infected cells for 50 h (see Note 5).
4. Recover cells by centrifugation in conical 500 mL tubes spun at 800 × *g* for 10 min.

5. Decant supernatant back into the culture flasks and kill virus by adding bleach to 10% final volume (see Note 6).
6. Resuspend pellet in 30 mL PBS to wash cells and transfer to a 50 mL conical tube.
7. Re-sediment cells in 50 mL tubes at 800 × *g* for 10 min. Decant supernatant carefully to avoid dislodging the cell pellet.
8. Freeze cell pellet in liquid nitrogen and store at −80°C (see Note 7).

3.2. Purification of HIS-Pak1 Recombinant Protein

1. Thaw the cell pellet on ice and conduct all subsequent steps at 4°C.
2. Resuspend the cell pellet derived from 1 L of culture with 16 mL of Lysis Buffer.
3. Sonicate gently using Branson tip sonifier in an ice bath and monitor lysis by microscopy (see Note 8).
4. Pellet the debris by centrifugation at 25,000 × *g* for 20 min. Collect the supernatant and add sodium azide to 0.02% to prevent bacterial growth.
5. Add 2 mL of Nickel-NTA beads (packed bead volume) prewashed into Washing Buffer (~20 mL) and rotate at 4°C for 2 h.
6. Pellet beads gently at 800 × *g* for 5 min and remove the depleted lysate (save in case some HIS-Pak1 remains in the lysate).
7. Wash the beads with 3 × 40 mL Washing Buffer.
8. Transfer to a disposable column for elution and elute with 10 sequential elutions of 2 mL Elution Buffer.
9. Save 5 μL of each elution and add 10 μL SDS-PAGE Sample Buffer. Boil for 4 min and load 10 μL for SDS-PAGE. A small sample of the eluted beads can be boiled in Sample Buffer and analyzed in parallel to assess the efficiency of elution.
10. Coomassie stain the gel. Pool HIS-Pak1 fractions of sufficient purity.
11. Determine the protein concentration of the pooled fractions using any method (e.g., Bradford, BCA).
12. Dialyze the pooled HIS-Pak1 against 1 L Pak Storage Buffer three times for 3 h each at 4°C.
13. Recover dialysate, add glycerol to 10%, aliquot, and snap freeze in liquid nitrogen.
14. Store at −80°C (see Note 9 regarding rapid freezing and thawing of purified proteins).

3.3. Expression/Purification of Cleaved Cdc42

During the expression and the purification steps, we recommend taking samples for SDS-PAGE analysis at the indicated steps in italics. These samples are useful for troubleshooting in the event of poor protein yield.

1. Set up an overnight culture of BL21 *E. coli* expressing GST-Cdc42 in 50 mL LB broth containing 100 μg/mL ampicillin (at 37°C).
2. In the morning, use the preculture to inoculate 950 mL of LB containing 100 μg/mL ampicillin. Monitor the growth of the culture by absorbance at 600 nm (A_{600}). Grow until A_{600} is in the range of 0.4–0.6 (relative to a sterile LB blank) at 37°C. Prior to induction, remove 10 μL sample of the culture and add 10 μL SDS-PAGE Sample Buffer. Boil for 4 min and load 10 μL for SDS-PAGE.
3. Add IPTG to 0.1 mM final in the 1 L culture to induce protein expression. Grow for 4–5 h. Remove 10 μL of the culture and add 10 μL SDS-PAGE Sample Buffer. Boil for 4 min and load 10 μL for SDS-PAGE. GST-Cdc42 should be strongly induced relative to the preinduction control.
4. Sediment the bacteria at 10,000×*g* for 20 min (4°C) and decant the supernatant.
5. Add 20 mL Lysis Buffer per cell pellet originating from a 1 L culture and resuspend well on ice.
6. Lyse bacteria by sonication using a Branson tip sonifier in an ice bath.
7. Pellet cell debris at 25,000×*g* for 30 min (4°C). Save 5 μL of the supernatant and add 10 μL SDS-PAGE Sample Buffer. Compare this sample with that from **step 3** above to confirm extraction of soluble protein from the total lysate.
8. Incubate the supernatant with 2 mL of glutathione agarose beads prewashed into Washing Buffer (~10 mL) per L of original culture volume.
9. Wash the beads extensively with PBS containing 1 mM $MgCl_2$ and 25 μM GDP (20 mL).
10. Estimate protein yield by Coomassie-stained SDS-PAGE analysis of a sample of the beads compared to known amounts of a standard protein (e.g., bovine serum albumin).
11. Add thrombin (amount based on the yield of GST-Cdc42 and the specific activity of the thrombin) and digest at room temperature for several hours. Assess cleavage efficacy by Coomassie-stained SDS-PAGE.
12. Collect the cleaved protein and rotate with benzamidine sepharose beads to remove the thrombin.
13. Collect the supernatant, conduct a protein assay to determine yield of cleaved protein, and add glycerol to 10% before snap freezing in liquid nitrogen in single-use aliquots.
14. Store at −80°C.

3.4. GTPγS Charging of Cdc42

1. Thaw rapidly at 37°C the required number of aliquots of cleaved Cdc42 and then transfer to ice.
2. Add GTPγS from a 10 mM stock to the Cdc42 to a final concentration of 200 μM.
3. Add EDTA to a final concentration of 2 mM.
4. Incubate for 20 min at 30°C.
5. Add $MgCl_2$ to 5 mM and transfer to ice.
6. Store on ice or freeze in liquid nitrogen.

3.5. Solution Preparation for Screening (Adequate for 12 Assay Plates)

1. Allow six compound plates to come to room temperature by incubation in a desiccator for an hour (see Note 10).
2. *Kinase solution*: Dilute Pakl in 1× Phospho Buffer to 50 μg/mL Pak1; ~35 mL of this solution is required (this includes the dead volume required to prime the MicroFill dispenser).
3. *MBP, Cdc42, and ATP solution*: Dilute MBP, GTPγS-charged Cdc42, and ATP into a single tube of 1× Phospho Buffer to obtain these final concentrations: MBP (0.32 mg/mL), GTPγS-charged Cdc42 (2.9 mg/mL), and ATP (21 μM). ~30 mL of this solution is required.
4. *Kinase-Glo*: Thaw and prepare Kinase-Glo reagents as directed by the manufacturer. 50 mL is required.

3.6. Screening

1. *Aliquot the Kinase solution into all 12 assay plates.* Using the MicroFill, dispense 8 μL of kinase to all screening columns of the 384-well assay plate (columns 2–23) (see Note 11). Store assay plates with lid at 4°C and flush MicroFill extensively with water.
2. *Pin transfer control compounds into control wells.* Transfer DMSO and staurosporine controls to the designated wells in columns 2 and 23 by pin transfer using a disposable polypropylene pin replicator from a DMSO stock plated in a 384-well plate. We recommend using 1 entire column each as replicate positive and negative control wells. To calibrate the concentration of inhibitor obtained by pin transfer, see Note 12.
3. *Pin transfer test compounds into assay plates.* Using the 384-pin replicator, transfer compounds from stock compound plates into assay plates and agitate gently to facilitate mixing.
4. *Aliquot MBP, Cdc42, and ATP solution into all 12 assay plates.* Dispense 7 μL of the mix containing MBP, Cdc42, and ATP solution into all screening columns (2–23) using the MicroFill. This step initiates the kinase reaction. The addition of MBP/Cdc42/ATP to each plate should be staggered by the amount of time required for the luminometer to read each plate so that all plates are incubated for precisely 2 h.

5. *Incubate plates at 30°C for 2 h.*
6. *Aliquot Kinase-Glo solution.* Dispense 10 μL of Kinase-Glo solution to columns 2–23 using the MicroFill and incubate for 10 min at room temperature.
7. *Read each assay plate using a luminometer according to the Kinase Glo protocol.*
8. *Data analysis.* Data is analyzed on a plate-by-plate basis in Microsoft Excel. Mean and standard deviation of the luminescence intensity of all negative control (DMSO only) wells are determined. Staurosporine control wells (positive controls) are examined to ensure quantitatively similar inhibition between plates. Wells containing test compounds that produce luminescence intensity greater than three standard deviations above the mean of control wells in the same plate are considered hits. Only compounds that hit in both plate replicates are considered further.

3.7. ^{32}P-ATP Kinase Assay (Secondary Screen)

Users of radioactive materials should be licensed and hold all required training certifications and permits.

1. Preheat a heating block to 30°C.
2. Label 1.5 mL locking-lid microcentrifuge tubes for each sample, including both negative (DMSO only) and positive controls (staurosporine).
3. Prepare a solution of 10 mM ATP containing 0.5 μCi [γ-^{32}P] ATP per μL.
4. For a 20 μL reaction volume, calculate the volume of stock solutions of Pak1, MBP, Cdc42-GTPγS, and test compound required for a final concentration of 500 nM Pak1, 10 μM MBP, 4 μM Cdc42-GTPγS, and 10 μM compound (see Note 13). Additional reaction components are 4 μL 5× Kinase Assay Buffer, 2 μL of the ATP solution prepared above, and water to bring the final volume to 20 μL.
5. Aliquot 5× Kinase Assay Buffer, water, and Pak1 into each reaction tube on ice.
6. Add individual inhibitors to each reaction tube and incubate for 5 min at room temperature.
7. Add Cdc42-GTPγS and MBP to each tube and incubate for an additional 5 min.
8. Start the reaction by adding 2 μL of the ATP solution and incubate for 10 min at 30°C in the heating block. If conducting multiple reactions, stagger the start times so that all tubes are incubated for precisely 10 min.
9. Stop the reaction by placing each microcentrifuge tube on dry ice.

10. Add 5 μL of 6× SDS-PAGE Sample Buffer to each microcentrifuge tube and boil for 5 min at 95°C (heating block) to stop the reaction and to prepare them for SDS-PAGE analysis.
11. Load 15 μL of each sample on a 15% SDS-PAGE gel.
12. Stop the SDS-PAGE gel before the dye front reaches the end of the gel to retain free [γ-^{32}P]ATP within the gel (which runs near the dye front).
13. Slice off the upper stacking gel and the bottom portion of the resolving gel below 10 kDa (containing free [γ-^{32}P]ATP) using a single-edged razor blade and discard as radioactive waste.
14. Coomassie stain the gel, destain it (dispose of staining and destain solutions as radioactive waste) (see Note 14).
15. Analyze incorporated MBP phosphorylation and Pak1 autophosphorylation by PhosphorImager analysis of incorporated ^{32}P in comparison with solvent control reactions (DMSO alone).
16. ATP-competitive inhibitors identified in the primary screen (using 10 μM ATP) are much less likely to exhibit significant inhibition under the secondary assay screen in the presence of 1 mM ATP.

4. Notes

1. Store cell culture media in the dark at 4°C. Warm medium to 27°C for use.
2. Other protease inhibitor cocktails may be substituted.
3. PMSF should be added just prior to lysis of bacteria, as it is unstable in aqueous solutions.
4. MBP can also be purified as described in ref. 13.
5. At 50 h, cells should show evidence of infection, swelling, and altered nuclear morphology.
6. After a 1 h incubation in the 10% bleach solution, inactivated virus can be disposed of safely.
7. Pellet can be stored >1 year at −80°C or can be used immediately without freezing.
8. 90% of the cells should be lysed before proceeding to the next step.
9. Because freeze/thaw cycles decrease protein stability, samples for frozen storage are best dispensed and prepared in single-use aliquots so that, once thawed, the protein solution will not need to be refrozen. For long-term storage (>1 month),

cryoprotectants such as glycerol can be added to a final concentration of 10%. Rapid freezing in liquid nitrogen is preferred prior to storage at −80°C. Concerning thawing of frozen aliquots, frozen protein solutions are rapidly thawed in a 37°C water bath and then immediately transferred to ice for use.

10. DMSO is highly hygroscopic and will absorb atmospheric water, resulting in compound dilution and, in some cases, precipitation.
11. Our compound library plates are formatted with compounds in columns 3–22. Columns 2 and 23 are used for control wells containing DMSO alone (negative control) or staurosporine (positive control). Columns 1 and 24 are not used to avoid plate edge effects.
12. To calibrate the pin transfer tool, use the pin transfer tool to transfer a known concentration of a fluorophore (e.g., fluorescein) dissolved in DMSO from a stock plate into DMSO-containing wells of a "mock" assay plate. Adjacent wells in the mock assay plate should contain a series of solutions of the same fluorophore dissolved in DMSO at carefully standardized concentration. Using a fluorimeter, establish a standard curve of fluorophore concentration versus fluorescence intensity. Using this standard curve, calculate the concentration of fluorophore in the pin-transferred wells. Based on the volume of DMSO in these wells and the concentration of fluorophore in the fluorophore stock plate, calculate the volume transferred by the pin tool.
13. The amount of DMSO in the final reaction should not exceed 1% to avoid nonspecific inhibition by the solvent.
14. Analyze the stained gel carefully for evidence of altered protein mobility. We have found that commercial compound collections can contain compounds that covalently modify or cause precipitation of proteins in a nonspecific way. These artifacts are easily visualized in the Coomassie-stained gel.

Acknowledgments

This work described here was supported by a W.W. Smith Foundation Award and an American Cancer Society Scholar Award to J.R.P. and by a National Institutes of Health (NIH) award RO1 GM083025 to J.R.P.

J.V. was supported by a grant from the Fondation pour la Recherche Médicale.

References

1. Bain J, McLauchlan H, Elliott M, Cohen P (2003) The specificities of protein kinase inhibitors: an update. Biochem J 371:199–204
2. Deacon SW, Beeser A, Fukui JA, Rennefahrt UE, Myers C, Chernoff J, Peterson JR (2008) An isoform-selective, small-molecule inhibitor targets the autoregulatory mechanism of p21-activated kinase. Chem Biol 15:322–331
3. Viaud J, Peterson JR (2009) An allosteric kinase inhibitor binds the p21-activated kinase autoregulatory domain covalently. Mol Cancer Ther 8:2559–2565
4. Eswaran J, Soundararajan M, Knapp S (2009) Targeting group II PAKs in cancer and metastasis. Cancer Metastasis Rev 28:209–217
5. Vadlamudi RK, Kumar R (2003) P21-activated kinases in human cancer. Cancer Metastasis Rev 22:385–393
6. Vadlamudi RK, Kumar R (2004) p21-activated kinase 1: an emerging therapeutic target. Cancer Treat Res 119:77–88
7. Kumar R, Gururaj AE, Barnes CJ (2006) p21-activated kinases in cancer. Nat Rev Cancer 6:459–471
8. Lei M, Lu W, Meng W, Parrini MC, Eck MJ, Mayer BJ, Harrison SC (2000) Structure of PAK1 in an autoinhibited conformation reveals a multistage activation switch. Cell 102: 387–397
9. Parrini MC, Lei M, Harrison SC, Mayer BJ (2002) Pak1 kinase homodimers are autoinhibited in trans and dissociated upon activation by Cdc42 and Rac1. Mol Cell 9:73–83
10. Strochlic TI, Viaud J, Rennefahrt UE, Anastassiadis T, Peterson JR (2010) Phosphoinositides are essential coactivators for p21-activated kinase 1. Mol Cell 40: 493–500
11. Gatti A, Huang Z, Tuazon PT, Traugh JA (1999) Multisite autophosphorylation of p21-activated protein kinase gamma-PAK as a function of activation. J Biol Chem 274: 8022–8028
12. Koresawa M, Okabe T (2004) High-throughput screening with quantitation of ATP consumption: a universal non-radioisotope, homogeneous assay for protein kinase. Assay Drug Dev Technol 2:153–160
13. Prowse CN, Hagopian JC, Cobb MH, Ahn NG, Lew J (2000) Catalytic reaction pathway for the mitogen-activated protein kinase ERK2. Biochemistry 39:6258–6266

Chapter 7

Using a Modified Yeast Two-Hybrid System to Screen for Chemical GEF Inhibitors

Anne Blangy and Philippe Fort

Abstract

GTPases of the Ras superfamily act as signaling switches, active when bound to GTP and inactive when bound to GDP. There is now considerable evidence that over-activation of Ras-like pathways participates in the development of many cancer types. In particular, GTPases of the Rho family control cell adhesion, survival, motility, and invasion, cell properties dysregulated in most cancer types. Rho activation is triggered by RhoGEFs, most of which form complexes with growth-factor receptors and initiate downstream Rho signaling pathways in response to extracellular clues. As such, RhoGEFs represent attractive targets to inhibit Rho pathways and may have interesting druggability for cancer therapeutics. Here we describe a procedure derived from the yeast two-hybrid system, in which activation of a mammalian Rho GTPase by its cognate RhoGEF is converted into variation in the yeast growth. The experimental design is thus suitable for identifying RhoGEF inhibitors and has been optimized for medium-throughput screening. The major advantages of this method lie in the direct monitoring of GEF activity in a living organism and the rapid detection of false positive hits.

Key words: Rho signaling, RhoGEF, Two-hybrid, Screening, Chemicals

1. Introduction

Over the past 20 years, Rho signaling pathways were shown to control basic cell properties (adhesion, polarity, migration, contraction, proliferation, and apoptosis), implicated in many normal and pathological processes, from cell movements in the developing embryo to common diseases, such as hypertension, cancer, or neurodegenerative diseases (1, 2).

Rho GTPases act as signaling switches that oscillate between inactive GDP-bound and active GTP-bound conformations. The activation step is promoted by RhoGEFs (guanine nucleotide exchange factors) of the Dbl and the CZH/DOCK families. Only once activated the GTPase specifically binds to downstream

Yi Zheng (ed.), *Rational Drug Design: Methods and Protocols*, Methods in Molecular Biology, vol. 928,
DOI 10.1007/978-1-62703-008-3_7, © Springer Science+Business Media New York 2012

effectors, thereby allowing the input signal to be converted into cellular outcomes (3).

In mammals, there are 20 Rho GTPases (4), 65 Dbl-related (5), and 11 DOCK-related proteins (6). Most RhoGEFs are high-molecular-weight proteins bound to membrane receptor complexes through the presence of specific domains (e.g., spectrin, SH3, FYVE, PH) in addition to the catalytic domain that promotes nucleotide exchange (DH for the Dbl family or DHR2 for the DOCK family). RhoGEFs therefore represent major cellular entry points whereby extracellular cues are converted into Rho signaling cascade activation.

In contrast to Ras proteins, whose activation in many tumor types results from somatic mutations that lock the GTPase in its active state (7), Rho hyperactivation in tumors was mostly found associated with over-expression of GTPases or RhoGEFs (8). Because of their high diversity and upstream position in Rho signaling cascades, RhoGEFs thus appear the most attractive targets for screening drugs that inhibit Rho pathways (2, 9).

Several approaches, classically used to monitor the catalytic activity of RhoGEFs, might be used for screening purposes: In vitro guanine nucleotide exchange assays using purified proteins or domains allow biochemical analysis of the exchange reaction (10). In vivo, the ectopic expression of an exchange factor induces the activation of endogenous GTPases, which can be monitored either directly by pull-down methods (11) or indirectly through analysis of their cellular activity. Nevertheless, both approaches suffer from major drawbacks: the purification of active GTPases and full-length RhoGEFs for in vitro exchange assays is frequently associated with insolubility and non-proper folding (12), whereas in vivo, data interpretation might be misleading due to cross talk between different Rho GTPases.

To overcome these limitations, we previously developed the Yeast Exchange Assay (YEA), a sensitive test based on a two-hybrid system in which Rho GTPase activation by a RhoGEF is monitored through interaction with a downstream effector (13). This assay was next updated and optimized for medium-throughput screens for RhoGEF inhibitors (14), which we validated by identifying a cell active inhibitor of the first exchange domain of the Trio multifunctional protein (15) and of the Dock5 exchange factor (16).

Theoretically, YEA strains can be engineered for screening inhibitors to any mammalian RhoGEF, even RhoGEFs for RhoA or Cdc42 whose homologues are expressed in the yeast *Saccharomyces cerevisiae* (13), and to any mammalian GEF activating GTPases of other Ras-like families like Ras, Rab, or Arf.

The experimental procedure described here was established using strains expressing wild-type or V12 mutant Rac1 and the kinectin Rho Binding Domain as two-hybrid partners (17), and the DHR2 domain of DOCK5 as a Rac1 RhoGEF (16).

2. Materials

2.1. Yeast Culture and Transformation

1. YAPD complete rich medium: 1% Yeast Extract, 2% Bacto-Peptone, 2% D-Glucose, 50 mg/L L-Adenine hemisulfate salt. For solid plates, add 2% Bacto agar (see Note 1).
2. YNB: 0.2% Yeast Nitrogen Base, 0.5% NH_4SO_4. For solid plates, add 2% Bacto agar (see Note 2).
3. SD minimal medium: YNB with 2% D-Glucose.
4. SD nutrient supplements: The following amino-acid supplements (100× stocks, see Note 3) should be added to the minimal SD medium, depending on the yeast strain and the selection markers on the plasmids used: 2 g/L L-Arginine HCl, 3 g/L L-Isoleucine, 3 g/L L-Methionine, 5 g/L L-Phenylalanine, 10 g/L L-Threonine, 2.5 g/L L-Histidine HCl monohydrate, 5 g/L L-Leucine, 2.5 g/L L-Adenine hemisulfate salt, 5 g/L L-Lysine HCl, 5 g/L L-Tryptophan, 2.5 g/L L-Uracil. For convenience, a 50× stock of all nutrients (50× AA) except ADE, LEU, and HIS can be prepared (see Note 5).
5. G418: Prepare 1,000× stock at 200 mg/mL (see Note 4).
6. 10× PEG: 50% Polyethylene glycol 3,350 (10×) (see Note 5).
7. 10× LiAc: 1 M Lithium Acetate at pH 7.5 (10×) (see Note 5).
8. 10× TE: 100 mM Tris–HCl pH 8, 10 mM EDTA (see Note 5).
9. 10 mg/mL Sonicated herring sperm DNA (see Note 6).
10. Sterile DMSO.
11. Z-buffer: 60 mM $Na_2HPO_4{\cdot}7H_2O$, 40 mM $NaH_2PO_4{\cdot}H_2O$, 10 mM KCl, 1 mM $MgSO_4$, 50 mM 2-mercaptoethanol, adjust pH to 7.0 if necessary (see Note 5).
12. 50 mg/mL X-Gal (50×) stock solution in sterile DMSO (see Note 7).
13. Liquid nitrogen.
14. 85–90 mm diameter nylon membranes and Whatman filter papers (see Note 8).
15. Flat bottom 96-well UV-Star plates from Greiner Bio-One (see Note 9).
16. V-bottom 96-well plates of 200 μL.
17. Gas-permeable adhesive seals.

2.2. Yeast Strains, DNA, Plasmids, and Oligonucleotides

1. TAT7 2-hybrid yeast: TAT7 (Mata, trp1, his3, leu2, ura3, ade2, LYS::(LexAop)4-HIS3, URA3::(LexAop)8-lacZ) or equivalent (see Note 10).
2. Strain to recover erg6 deletion: Y00568 (erg6Δ, isogenic to BY4741 with YML008c::kanMX4) available from EUROSCARF (Frankfurt, Germany).

3. The wild-type uncaaxed GTPase of interest fused to the LexA DNA binding domain in a yeast 2-hybrid plasmid (pLex) and its effector fused to the GAL4 activation domain in a yeast 2-hybrid plasmid (pGAD). We, respectively, use pBTM116 with a TRP1 selection marker and pGAD with an LEU2 selection marker (see Note 11).
4. Yeast integrative vector with the LEU2 selection marker that expresses both the wild-type uncaaxed GTPase fused to the LexA DNA-binding domain and the effector fused to the GAL4 activation domain. You can for instance replace the 2μ replication origin of the pGAD plasmid with the promoter and the LexA-fused GTPase of the pLex plasmid.
5. Yeast vector multicopy plasmid with ADE2 selection marker such as pRs422 and expressing the exchange factor (GEF) of interest under a yeast promoter, ideally with a myc tag or so (see Note 12).
6. Primers to amplify and verify the insertion of the ERG6::KAN cassette:

   UPERG6: 5′-GCTGTTGCCGATAACTTCTTCATTGC-3′

   UPUPERG6: CGAAGATTGGTGAGAAACCTC-3′

   DOWNERG6: CTGATAGAAAATACTGGTCGTTTGCCACG-3′

   DOWNDOWNERG6:5′-GTCAATACGTTTGTATGCAGTG-3′

   KAN787: 5′-TTGCCATCCTATGGAACTGC-3′

   KAN1023: 5′-ACGACTGAATCCGGTGAGAA-3′ (see Note 13)
7. 456–600 μm glass beads.
8. 1:1 Phenol–chloroform.
9. 100 and 70% ethanol.
10. Lysis buffer: 10 mM Tris–HCl pH 8, 1 mM EDTA, 100 mM NaCl, 1% SDS, 2% Triton X-100.

3. Methods

3.1. Transforming Yeasts

1. Streak the TAT7 yeast strain on a YAPD plate and incubate for 2 days at 30°C.
2. Inoculate 50 mL YAPD with a single colony and grow overnight at 30°C with shaking.
3. All the following steps are performed at room temperature. Centrifuge cells in a 50 mL sterile falcon tube at 500 × g for 5 min.
4. Resuspend pelleted cells in 50 mL sterile water and centrifuge again as above.

5. Redo step 4 with 10 mL of water.
6. Resuspend pelleted cells in 10 mL 1× LiAc/TE, freshly prepared from the 10× stock solutions. Centrifuge again as above (see Note 14).
7. Resuspend pelleted cells in 250 μL 1× LiAc/TE. Yeasts are now competent for transformation. They can be stored at −80°C in 100 μL aliquots after addition of sterile glycerol to 10% (see Note 15).
8. Combine 1 μg of the plasmids to be transformed in yeast to sonicated salmon sperm DNA q.s. 5 μg of DNA in a 1.5 mL microtube. Vortex (see Note 16).
9. Add 50 μL of competent cells and mix by pipetting up and down.
10. Add 300 μL of 40% PEG in 1× LiAc/TE. Vortex vigorously to perfectly mix cells in the viscous PEG.
11. Leave for 30 min in a 30°C incubator.
12. Heat shock in a 42°C water bath for 15 min.
13. Spin cells for 10 s in a microfuge.
14. Resuspend in 150 μL of sterile water.
15. Plate on the appropriate selection medium and grow for at least 2 days at 30°C to obtain colonies.
16. Re-streak four colonies on a new selection plate for further use.
17. For long-term storage, grow cells in liquid selection medium overnight at 30°C. Refresh on ice and add sterile glycerol to 10%. Store in 1 mL aliquots at −80°C.

3.2. Extent of GTPase Activation by the Exchange Factor in the Yeast Reporter Assay: Qualitative X-gal Staining

1. According to the above protocol, prepare TAT7 transformed with the yeast 2-hybrid plasmids expressing the wild-type uncaaxed GTPase of interest fused to the LexA DNA-binding domain and the effector fused to the GAL4 activation domain, as well as with the centromeric plasmid empty (−GEF) or expressing (+GEF) the exchange factor.
2. Re-streak 4–6 colonies of each transformation on a selective plate and grow for 1–2 days at 30°C.
3. Place a nylon disk onto a selective plate, streak the colonies again, and grow for 1–2 days at 30°C.
4. Freeze the nylon disk by dipping for 5 s into liquid nitrogen and place on the bench to warm up.
5. Place 0.5–1 mL of Z-buffer containing 1× X-Gal inside the cover of a Petri dish and place a filter paper on top. The paper should absorb all the buffer and be soaked evenly.
6. Place the nylon membrane on top of the soaked filter paper and put the bottom of the Petri dish on top, as a cover.

7. Incubate at 37°C for a few minutes to a few hours. Yeast expressing ß-galactosidase, indicating the activation of the GTPase by the exchange factor, will progressively turn bluish. The activation of the GTPase by the GEF should result in faster development of the color in the GEF expressing clones (see Note 17).
8. Stop the reaction when difference in color intensity is clear between +GEF and −GEF clones. For this, place the membrane at room temperature on a 300-μL drop of 10% acetic acid on a plastic foil and leave until it is dry. Store posted on Whatman filters wrapped in Saran-wrap.

3.3. GTPase Activation by the GEF in the YEA System: Quantitative Growth in Histidine Selection Medium

1. Pick one clone from each of the two transformations (+GEF and −GEF) above and of untransformed TAT7 yeast. Inoculate in 50 mL SD medium complemented for the appropriate nutrients and lacking the nutrients to select for the plasmids when necessary (here LEU and ADE). Grow for 24 h (until late afternoon) at 30°C in a shaker.
2. Centrifuge cells in a 50 mL sterile falcon tube at 500 × *g* for 5 min and resuspend the pellet in 50 mL YNB. Perform this procedure twice (see Note 18).
3. Measure OD_{600} of each of the three cultures (TAT7, +GEF, and −GEF). For each culture, start two new 50 mL cultures at $OD_{600} = 0.2$ in nutrient-complemented SD medium as in 1, lacking histidine (−HIS) or not (+HIS). Grow at 30°C in a shaker. Take OD_{600} at T_0 for each of the six cultures and then at $T = 2$ h (see Table 1).
4. The following days, measure OD_{600} of the cultures every 2–3 h.
5. Do the same the following day.
6. Calculate yeast growth rates from the linear parts of the curves (see Fig. 1). For each time point, determine the four growth rates (see Table 1): GR1 (−GEF/+HIS), GR2 (−GEF/−HIS), GR3 (+GEF/+HIS), and GR4 (+GEF/−HIS) (see Note 17).
   (a) GR2/GR1 indicates the activation of the GTPase by endogenous yeast proteins. Ideally, it should be close to zero.
   (b) GR3/GR1 should be very close to 1. Otherwise, it indicates an irrelevant effect of the expressed GEF on yeast growth.
   (c) The ratio GR4/GR2 measures the activation of the HIS3 reporter gene induced by the expression of the GEF, i.e., activation of the GTPase by the GEF.

Table 1
Growth rate ratios and R_{GEF} as defined in Subheadings 3.3 and 3.5

	Growth times (h)		
Growth rate (GR)[a]	15	18	23
GR1 (−GEF/+HIS)	0.04554	0.04569	0.04203
GR2 (−GEF/−HIS)	0.01967	0.01931	0.0188
GR3 (+GEF/+HIS)	0.04409	0.04441	0.04166
GR4 (+GEF/−HIS)	0.03163	0.03302	0.03433
GR2/GR1	0.43	0.42	0.45
GR3/GR1	0.97	0.97	0.99
GR4/GR3	1.61	1.71	1.83
GR2/GR4 (R_{GEF})	0.62	0.58	0.55

Yeast strains are described in the legend of Fig. 1
[a]GR1-4 values are plotted in Fig. 1

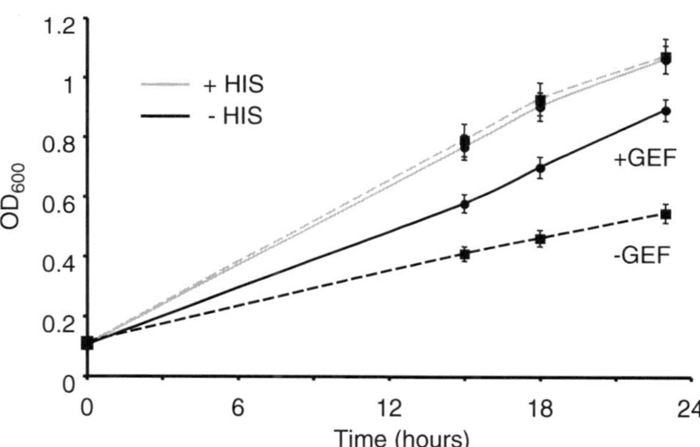

Fig. 1. Faster growth in histidine selection medium of yeasts expressing the exchange factor. TAT7 erg6Δ yeasts were transformed with integration plasmids expressing wild-type Rac1 fused to LexA DNA-binding domain and the Rac-binding domain of kinectin (K66, [17]) fused to GAL4 activation domain and with a multicopy vector expressing (+GEF) or not (−GEF) the DHR2 exchange domain of Dock5 [18]. They were grown in 96-well plates in SD-LEU-ADE with or without histidine (+HIS or −HIS), according to the procedure described in Subheading 3.3. OD_{600} of the wells was measured at 15, 18, and 23 h after yeast dilution. Measurement values are shown in Table 1. The graph shows average and standard deviation of 12 wells at each time point.

(d) GR2/GR4 (R_{GEF}) is thus the highest reduction in yeast growth you will be working with when testing inhibitors of the GEF, GR2 corresponding to a complete inhibition of GEF activity. R_{GEF} should be below 0.8.

3.4. Disruption of erg6 in TAT7

1. Grow Y00568 in 2 mL YAPD overnight at 30°C to prepare the DNA.
2. Transfer culture into 1.5 mL microtubes and spin for 10 s in a microfuge at maximum speed.
3. Rinse pellet in 1 mL sterile water and spin again.
4. Resuspend pellet in 0.2 mL of lysis buffer and add about 0.3 g of glass beads. *Note*: Check the volume it represents in your microtube. Add 0.2 mL of phenol–chloroform.
5. Vortex the tube at top speed for 2 min.
6. Add 0.2 mL of 1× TE and vortex to mix.
7. Spin for 5 min in a microfuge at maximum speed.
8. Recover the supernatant without touching the interface.
9. Mix with 2 volumes of 100% ethanol.
10. Spin for 5 min in a microfuge at maximum speed to pellet the DNA.
11. Remove the supernatant and rinse the pellet with 1 mL of 70% ethanol.
12. Spin for 1 min in a microfuge at maximum speed.
13. Discard ethanol and leave the pellet to dry in the open tube.
14. Dissolve the pellet in 0.1 mL 1× TE.
15. Amplify the yeast DNA with UPERG6 and DOWNERG6 using a high-fidelity Taq polymerase setting the annealing temperature to 53°C. The ERG6::KAN fragment is 2.1 kb long whereas the PCR of the non-disrupted erg6 gene using the same primers would give a 1.5 kb fragment (see Note 19).
16. Purify the ERG6::KAN fragment on an agarose gel.
17. Transform TAT7 with 1 μg fragment according to the protocol above. Plate on YAPD containing 200 μg/mL G418 and leave for 3–4 days at 30°C.
18. Re-streak 8–10 colonies on a new YAPD plate containing G418.
19. Prepare yeast DNA for direct PCR screening of the insertion by resuspending a small amount of yeast (1–2 μL) in 20 μL of water in a microfuge tube. Microwave for 2 min at maximum power. Spin down and use 2–3 μL per PCR (see Note 20).
20. Verify the insertion by PCR using the three following primer sets: UPUPERG6 and DOWNDOWNERG6 (2.1 kb), KAN787 and DOWNDOWNERG6 (0.6 kb), and UPUPERG6 and KAN1023 (1.3 kb) (see Note 21).

3.5. Preparation of Yeast Strains for Screening Inhibitors of GEFs

1. Transform the TAT7 erg6Δ yeast with the integration plasmid expressing the GTPase and its effector and with the vector expressing (**test** strain) or not (**control** strain) the GEF.
2. Allow the two yeast strains obtained as described in Subheading 2.2 to grow in SD medium complemented with 1× AA (SD-LEU-ADE) with (+HIS) and without (−HIS) histidine. Verify that the characteristics of the growth rates GR1-GR4 and R_{GEF} are appropriate according to Subheading 3.3.
3. Test the sensitivity to DMSO of the two yeast strains in a yeast growth assay in SD medium complemented with 1× AA and histidine. You can usually go up to 1% without affecting yeast growth.
4. Grow a 15 mL pre-culture of the control and test strains, streaked 2–3 days before on a SD-LEU-ADE+HIS plate, in SD-LEU-ADE+HIS medium overnight at 30°C.
5. The next morning, dilute the cultures to $OD_{600} = 0.2$ in 30 mL SD-LA+HIS medium.
6. When OD_{600} reaches 0.4, centrifuge cells in a 50 mL sterile falcon tube at 500×g for 5 min and resuspend the pellet in 50 mL cold YNB.
7. Centrifuge cells in a 50 mL sterile falcon tube at 500×g for 5 min and resuspend the pellet in 30 mL cold YNB.
8. Dilute cells to $OD_{600} = 0.25$ in cold YNB. You need about 4 mL for one 96-well plate (see Note 22).
9. Fill up a 96-well Greiner UV-Star plates with the following:
   (a) Line 1, 3: 160 μL SD-LEU-ADE-HIS, 10 μL of 20% DMSO, and 40 μL of test yeast.
   (b) Line 2, 4: 160 μL SD-LEU-ADE+HIS, 10 μL of 20% DMSO, and 40 μL of test yeast.
   (c) Line 5, 7: 160 μL SD-LEU-ADE-HIS, 10 μL of 20% DMSO, and 40 μL of control yeast.
   (d) Line 6, 8: 160 μL SD-LEU-ADE+HIS, 10 μL of 20% DMSO, and 40 μL of control yeast.
10. Close with a gas-permeable adhesive seal and leave the plate for yeast sedimentation for 1 h on the bench.
11. Measure OD_{600} at T_0 of all wells after removing the seal of the plate. Take care not to shake the plates. Put the seal back on the plates and leave on bench at room temperature overnight.
12. The next day, measure OD_{600} around $T = 14$, 18, and 24 h.
13. Plot average and SD OD_{600} of the four time points.
14. Verify that the growth curves from lines 2 and 4 are identical (GR1 and GR3).

15. Calculate the average GR ratios defined in Subheading 3.3 in the absence and presence of the GEF or HIS, and in particular R_{GEF}:
    (a) $[\mathrm{OD}_{600}(T)-\mathrm{OD}_{600}(T_0)]^{-\mathrm{GEF}}/[\mathrm{OD}_{600}(T)-\mathrm{OD}_{600}(T_0)]^{+\mathrm{GEF}}$ (Fig. 1). Proceed to Subheading 3.6 if ratios are correct.

3.6. Proceeding with the Screening

Using a pipetting robot, you can consider handling 12 96-well plates of compounds at a time. The step-by-step procedure is illustrated in Fig. 2.

1. On day −6 or −5 before the screening, re-streak the test and control strains on an SD medium plate complemented with 1× AA (SD-LEU-ADE+HIS).
2. On day −3 before the screening, prepare freshly complemented growth media (500 mL of SD-LEU-ADE+HIS and 500 mL of SD-LEU-ADE-HIS). Start a 15 mL culture of the control strain and two 15 mL cultures of the test strain.

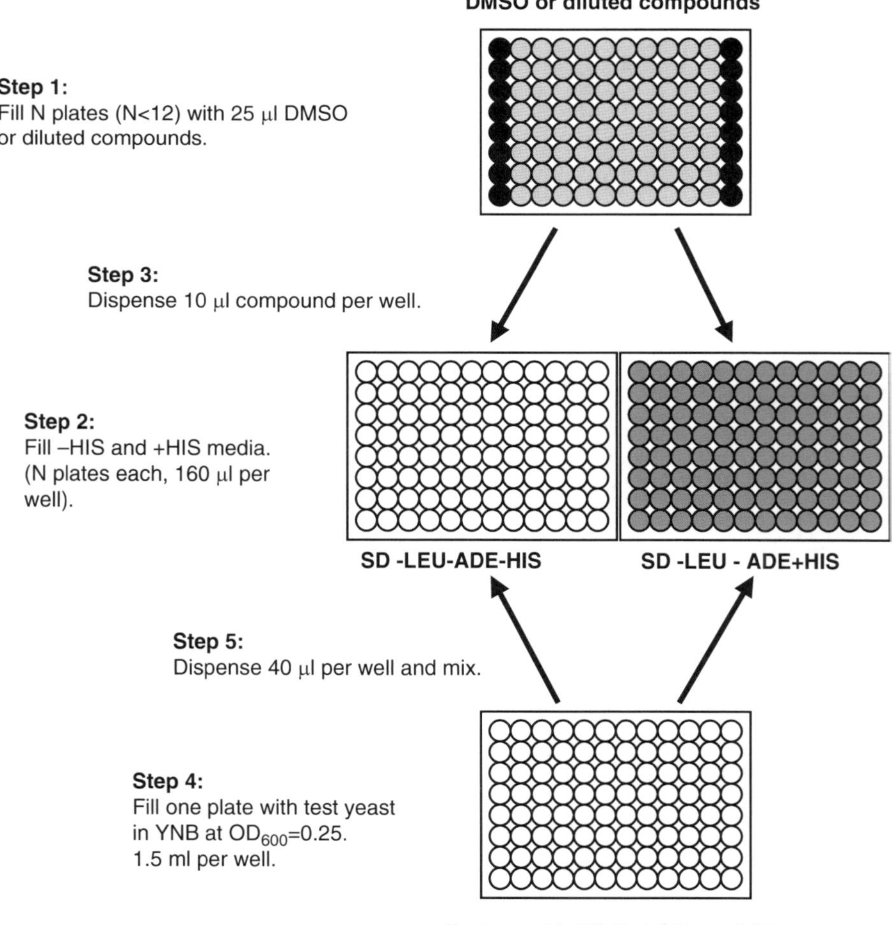

Fig. 2. Step-by-step scheme of the screening procedure. All steps are fully described in Subheading 3.6.

3. On days −2 and −1 before the screening, perform a growth assay as described in Subheading 3.5 to determine the average R_{GEF} from the three time points. Keep the second culture of test strain exponentially growing by diluting it on day −2 to $OD_{600}=0.2$ in 15 mL of the freshly prepared SD-LEU-ADE-HIS medium above (see Note 22).
4. On day −1 before the screening, start a 75 mL pre-culture of the test strain in the freshly prepared SD-LEU-ADE-HIS medium above.
5. The morning of the following day dilute the test strain culture in 150 mL of SD-LEU-ADE-HIS to $OD_{600}=0.2$. Grow until $OD_{600}=0.4$.
6. Meanwhile, dilute 5 μL chemical compounds (initially in 100% DMSO) in 20 μL sterile water. Dispatch each compound in 96-well plates with columns 1 and 12 of the plate containing 25 μL of 20% DMSO.
7. For each 96-well plate of compounds, prepare two Greiner UV star plates, one labeled +HIS and the other labeled −HIS.
8. Fill up wells of −HIS and +HIS plates with 160 μL of SD-LEU-ADE-HIS and SD-LEU-ADE+HIS, respectively.
9. Transfer 10 μL of the diluted compounds in one −HIS and in one +HIS plate.
10. When the yeast culture reaches $OD_{600}=0.4$, pellet the yeasts in 50 mL falcon tubes and rinse each pellet with 50 mL cold YNB.
11. Resuspend each pellet in 40 mL cold YNB and pool suspensions in a sterile flask. Adjust OD_{600} to 0.25 with cold YNB and dispense 1.5 mL in a 2 mL Deep Well 96-well plate; you need about 150 mL when starting from 12 96-well plates of compounds.
12. Add 40 μL of the test yeast strain diluted at $OD_{600}=0.25$ in YNB in each well. Mix well by slowly pipetting up and down five times. Close with gas-permeable adhesive seals (see Note 23).
13. Leave plates on bench for 1 h to allow yeasts to settle down.
14. Measure OD_{600} after removing the seals. It should be around 0.05. This is T_0.
15. Put the seals back on the plates and leave them at room temperature on the bench overnight. The next day, take OD_{600} at $T=14$, 18, and 22 h.

3.7. Interpretation of the Results and Selection of Potential Inhibitors

1. In each plate and for each time point, calculate the ratio between yeast growth rate with and without histidine in the presence of the compound (Cp):
2. $R_{Cp}=[OD_{600}(T)-OD_{600}(T_0)]^{-HIS+Cp}/[OD_{600}(T)-OD_{600}(T_0)]^{+HIS+Cp}$.

3. In each plate and for each time point, calculate from columns 1 and 12 the average and standard deviation of the ratio between yeast growth rate with and without histidine in the presence of DMSO:

4. $R_{DMSO} = [OD_{600}(T) - OD_{600}(T_0)]^{-HIS+DMSO} / [OD_{600}(T) - OD_{600}(T_0)]^{+HIS+DMSO}$.

5. Eliminate all the compounds for which $[OD_{600}(T) - OD_{600}(T_0)]^{+HIS+Cp}$ is less than 50% the average $[OD_{600}(T) - OD_{600}(T_0)]^{+HIS+DMSO}$ of the 16 wells in DMSO of the plate. These compounds are toxic for yeast growth.

6. Make a bar graph by plotting the average R_{DMSO} with standard deviation and all the 80 R_{Cp} values.

7. Select all the compounds for which R_{Cp}/R_{DMSO} is below the cutoff. We usually select compounds inhibiting growth by at least 50%, i.e., the average R_{Cp}/R_{DMSO} ratio from the three time points is below $(R_{DMSO} + R_{GEF})/2$ (Fig. 3).

8. Pool all the selected compounds in one 96-well plate and retest them as in Subheading 3.6.

9. Perform appropriate secondary validation of the hit compounds. Since inhibition at any step downstream of the GEF-activated GTPase is expected to produce same effects on yeast growth as GEF inhibition, a critical validation is to test compounds on growth of two-hybrid yeast strains expressing the LexA DNA-binding domain fused to a constitutively active GTPase mutant instead of the wild-type GTPase. Compounds that inhibit growth should be disregarded.

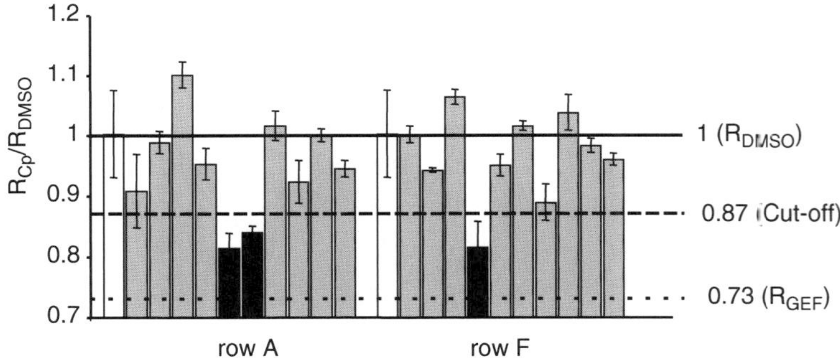

Fig. 3. Example of a screening result. Bars represent the average and SD R_{Cp}/R_{DMSO} values at the three time points for single compounds (*filled bars*) or control DMSO (*open bars*) from columns 1 or 12 of a 96-well plate (*black circles* in Fig. 2). Using the procedure defined in Subheading 3.5, we found that R_{DMSO} was close to 1 (*plain line*) and R_{GEF} for the exchange domain of DOCK5 was 0.73 (*dotted line*). We thus used a cutoff equal to 0.87 (*dashed line*), corresponding to the theoretical 50% inhibition of the GEF activity, i.e., $(R_{DMSO} + R_{GEF})/2$. Compounds with average R_{Cp}/R_{DMSO} ratios below the cutoff (*black bars*) were selected as potential inhibitors of Dock5, examplified here for two rows of the plate (A and F).

4. Notes

1. To prepare YAPD, dissolve Yeast Extract and Bacto-Peptone in H_2O; add the Bacto agar for solid plates. Autoclave at 120°C for 20 min. Let the medium cool at 50–60°C and complement with glucose and adenine. For long-term storage, keep autoclaved YAPD at room temperature un-complemented. It can be easily melt in a microwave for preparing plates. Prepare a separate 20% glucose solution (1 M) and autoclave as above. Store at room temperature. Store complemented medium plates at 4°C.
2. Add all components and autoclave as above. You can store YNB at room temperature.
3. Amino acid solutions should be filtered on 0.2 μm for sterilization. Store at 4°C. Tryptophan must be stored in the dark; you can wrap the tube with aluminum.
4. Filter on 0.2 μm for sterilization and store at −20°C.
5. Autoclave at 120°C for 20 min and store at room temperature.
6. Start from ready-to-use solutions rather than lyophilized DNA for better transformation efficiency.
7. Store at −20°C.
8. Nylon membranes are easier to handle after freezing in liquid nitrogen; nitrocellulose membranes tend to break into pieces.
9. We use the reference 736–0231 from Greiner Bio-One that gives homogeneous OD_{600} measurements across wells. In many 96-well plates, we found that yeast tends to aggregate.
10. This strain derived from L40 2 hybrid yeast strain has a very low background growth in −HIS medium and does not require 3-amino-1,2,4,-triazole (3-AT). If using another strain such as Y187, you will have to optimize the concentration of 3-AT in your assays.
11. When using TAT7, you must use the LexA DNA-binding domain fusions. If using another strain such as Y187 you may have to use the GAL4 DNA-binding domain.
12. You can also use a centromeric or an integrative plasmid, depending on the level of expression of the GEF needed to obtain sufficient GTPase activation.
13. Yeast strain devoid of ERG6 is unable to grow on tryptophan-depleted medium. You therefore cannot use plasmids with a trp1 selection marker.
14. The solution must be prepared fresh by mixing 10× PEG, 10× LiAc, and 10× TE. Mix well by vortexing the solution extensively.

15. Glycerol does not affect significantly plasmid transformation in yeast. You can store competent yeasts already bearing a plasmid.
16. Keep the volume below 10 μL.
17. If the color develops quickly in both types of clones or if GR2/GR6 is above 1 and GR2/GR1 above 0.8, that means there is a strong background activation of the GTPase in yeast. You may try to use the GTPase with its CAAX box, introduce thermosensitive mutations in the yeast activators of Rho GTPases (Cdc24 if you are working with a Cdc42 activator, ROM1 and ROM2 if you are looking at a RhoA GEF), or use a closely related Rho GTPase (RhoC instead of RhoA). If the color develops very slowly in both types of clones or if GR4/GR3 is very low (far from 1 and close to GR2/GR1) and GR4/GR2 is low (close to 1), the GTPase is not efficiently activated by the GEF. You may try other GEF constructs. If you are using another yeast strain, you need to measure the background yeast growth in the absence of histidine. You may need to optimize the difference in growth rates of your strain between +HIS and –HIS by adding 3-AT in the cultures to decrease GR2/GR1. This is not necessary when using TAT7.
18. Washing the yeast pellet before inoculation is very important to remove all trace of histidine.
19. We use the Expand Taq polymerase from Roche. You may have to test various quantities of DNA (0.1–5 μL). If you have no amplification, re-extract with phenol–chloroform and precipitate the DNA again.
20. Pick yeast from the plate with a pipetman tip. Do not use toothpicks; most of them are soaked with fixatives such as formaldehyde that inhibit the PCR reaction.
21. UPUPERG6 and DOWNDOWNERG6 are in the ERG6 gene, upstream and downstream the primers used to amplify the disruption cassette from Y00568. Use TAT7 and Y00568 DNA as controls for the size of disrupted and undisrupted ERG6.
22. This procedure leads to actively growing cells. Yeasts should not stop growing after dilution. It is very important for short-term growth assays. It is also important to prepare your medium fresh and perform the growth tests to make sure that R_{GEF} is below 0.8. It is important to start from exponentially growing cells and not to start with OD_{600} below 0.05; otherwise it will take hours for the cells to start growing again. It is also important not to start with OD_{600} above 0.1 because yeasts tend to aggregate when their concentration is too high and the window of linear increase of the OD would be too short.
23. If using a pipetting robot, you may dispatch the cells in 2 mL Deep Well 96-well plates. You should add a mixing step before pipetting the 40 μL to homogenize the solution.

Acknowledgments

We are indebted to Pauline Larrousse for technical support. This work was funded by institutional grants from CNRS and Montpellier Universities, and by contracts from ANR (ANR-06-PCVI-0024), ARC (n° 1048), FRM (DVO20081013473), and Arthritis Fondation Courtin (OPT30713).

References

1. Shirai H, Autieri M, Eguchi S (2007) Small GTP-binding proteins and mitogen-activated protein kinases as promising therapeutic targets of vascular remodeling. Curr Opin Nephrol Hypertens 16:111–5
2. Fritz G, Just I, Kaina B (1999) Rho GTPases are over-expressed in human tumors. Int J Cancer 81:682–7
3. Bosco EE, Mulloy JC, Zheng Y (2009) Rac1 GTPase: a "Rac" of all trades. Cell Mol Life Sci 66:370–4
4. Boureux A, Vignal E, Faure S, Fort P (2007) Evolution of the Rho family of ras-like GTPases in eukaryotes. Mol Biol Evol 24:203–16
5. Rossman KL, Der CJ, Sondek J (2005) GEF means go: turning on RHO GTPases with guanine nucleotide-exchange factors. Nat Rev Mol Cell Biol 6:167–80
6. Meller N, Merlot S, Guda C (2005) CZH proteins: a new family of Rho-GEFs. J Cell Sci 118:4937–46
7. Diaz-Flores E, Shannon K (2007) Targeting oncogenic Ras. Genes Dev 21:1989–92
8. Vega FM, Ridley AJ (2008) Rho GTPases in cancer cell biology. FEBS Lett 582: 2093–101
9. Lazer G, Katzav S (2010) Guanine nucleotide exchange factors for RhoGTPases: Good therapeutic targets for cancer therapy? Cell Signal 23:969–79
10. Leonard DA, Evans T, Hart M, Cerione RA, Manor D (1994) Investigation of the GTP-binding/GTPase cycle of Cdc42Hs using fluorescence spectroscopy. Biochemistry 33: 12323–8
11. Sander EE, van Delft S, ten Klooster JP, Reid T, van der Kammen RA, Michiels F, Collard JG (1998) Matrix-dependent Tiam1/Rac signaling in epithelial cells promotes either cell-cell adhesion or cell migration and is regulated by phosphatidylinositol 3-kinase. J Cell Biol 143:1385–98
12. Zheng Y, Hart MJ, Cerione RA (1995) Guanine nucleotide exchange catalyzed by dbl oncogene product. Methods Enzymol 256:77–84
13. De Toledo M, Colombo K, Nagase T, Ohara O, Fort P, Blangy A (2000) The yeast exchange assay, a new complementary method to screen for Dbl-like protein specificity: identification of a novel RhoA exchange factor. FEBS Lett 480:287–92
14. Blangy A, Bouquier N, Gauthier-Rouviere C, Schmidt S, Debant A, Leonetti JP, Fort P (2006) Identification of TRIO-GEFD1 chemical inhibitors using the yeast exchange assay. Biol Cell 98:511–22
15. Bouquier N, Vignal E, Charrasse S, Weill M, Schmidt S, Leonetti JP, Blangy A, Fort P (2009) A cell active chemical GEF inhibitor selectively targets the Trio/RhoG/Rac1 signaling pathway. Chem Biol 16:657–66
16. Vives V, Laurin M, Cres G, Larrousse P, Morichaud Z, Noel D, Cote JF, Blangy A (2011) The Rac1 exchange factor Dock5 is essential for bone resorption by osteoclasts. J Bone Miner Res 26(5):1099–110
17. Vignal E, Blangy A, Martin M, Gauthier-Rouviere C, Fort P (2001) Kinectin is a key effector of RhoG microtubule-dependent cellular activity. Mol Cell Biol 21:8022–34

Chapter 8

Random Mutagenesis of Peptide Aptamers as an Optimization Strategy for Inhibitor Screening

Nathalie Bouquier, Sylvie Fromont, Anne Debant, and Susanne Schmidt

Abstract

Accumulating work over the past decade has shown that peptide aptamer screening represents a valid strategy for inhibitor identification that can be applied to a variety of different targets. Because of the screening method in cells and the highly combinatorial libraries available, this approach yields rapidly highly specific candidate inhibitors. Once a hit peptide has been identified, its interaction strength and affinity towards its target protein can be optimized even more, in order to increase its inhibition efficiency when subsequently applied in vivo. A condition to a successful optimization is that gain of inhibition strength should not result in loss of specificity.

Here we present a simple method for peptide aptamer optimization, which can be achieved by PCR-based random mutagenesis combined with a selection screen in yeast using a strong selective drug. The rationale of this approach, which has proven valid and efficient, is that stronger interaction in yeast will also lead to stronger inhibition. Our optimization method is effective, without loss of specificity, which is of a great importance for the discovery of inhibitors that target specific protein–protein interactions.

Key words: Peptide aptamers, PCR-based mutagenesis, Yeast two-hybrid, Fluorescent GEF assay, Inhibitor screening

1. Introduction

Peptide aptamers are short peptides, generally composed of 20 random amino acids, which are conformationally constrained by a scaffold protein such as bacterial Thioredoxin A (TrxA), and which are commonly used as disrupters of protein–protein interactions (1). They have been widely used over the past decade as inhibitors of various intracellular targets, including cell cycle control and cell

Yi Zheng (ed.), *Rational Drug Design: Methods and Protocols*, Methods in Molecular Biology, vol. 928, DOI 10.1007/978-1-62703-008-3_8, © Springer Science+Business Media New York 2012

survival proteins, and also as antiviral and antibacterial agents (2, 3). They have also been applied to dissect and inhibit intracellular signaling pathways that regulate cellular proliferation, such as EGFR or Ras signaling (4). Our lab has used this technology to specifically target one of the Rho guanine nucleotide factor (RhoGEF) domains of Trio, which activates the small monomeric GTPase RhoA (5, 6). We initially isolated the peptide aptamer TRIPα as an inhibitor of Trio, using a classical two-hybrid screening (5). Subsequently, using the optimization strategy described in this chapter, we were able to convert this original aptamer into an even more potent inhibitor of Tgat, the oncogenic isoform of Trio, which activates RhoA and which has been isolated from patients with Acute T-cell Leukemia (ATL) (7). This novel TRIPα-derived peptide targets Tgat GEF activity in vitro in a highly specific manner and strongly reduces its oncogenic properties in vivo, most remarkably by decreasing foci formation and tumor development in nude mice (6). Interestingly, although TRIPα and the TRIP-like peptides were initially selected as inserted in the TrxA scaffold, they remained equally active as linear peptides.

Peptide aptamers present interesting advantages over other classes of inhibitory molecules, mainly because of their simple design and their high degree of binding specificity, which enables them to discriminate between closely related proteins within a functional family. As compared to an approach by complete knockdown of the protein (such as siRNA), peptide aptamers recognize selectively a single domain of a protein, thus interfering with one specific function, without affecting others. This is of great advantage when targeting complex proteins harboring numerous domains. But most remarkably, these unbiased and highly combinatorial proteins are screened and designed to function inside living cells.

High selectivity is clearly the major advantage of the peptide aptamer technology. We show here that inhibition strength may even be enhanced, which is of particular importance for the use of the peptides in in vivo systems, as demonstrated by Bouquier and colleagues (6). In this chapter, we present a simple method for peptide aptamer optimization, i.e., increased inhibition potential towards the target protein, which can be achieved by random PCR-based mutagenesis combined with a selection screen in yeast using a strong selective drug (Fig. 1). The rationale of this approach, which has proven valid and efficient, is that stronger interaction in yeast will also lead to stronger inhibition. The peptide aptamer isolated in the initial screen and selected for its specific inhibition of the target protein, is used as a template for PCR-based random mutagenesis, in order to create a new peptide aptamer library derived from this peptide, which can then be screened for stronger interaction with the bait.

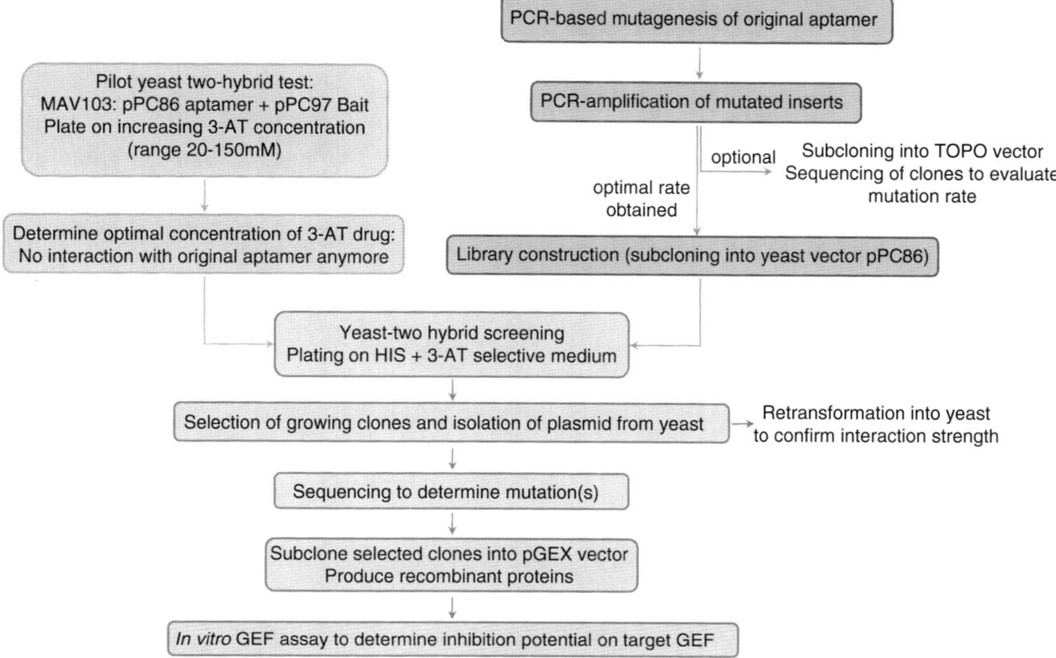

Fig. 1. Flow chart of the peptide aptamer optimization protocol.

PCR is performed using specific conditions that vary from the classical PCR, in order to maximize misincorporation of base pairs (8). High concentrations of $MgCl_2$ and $MnCl_2$ are used to stabilize noncomplementary base pairs and to diminish the template specificity of the Taq DNA polymerase, respectively. An excess of dCTP and dTTP (versus dATP and dGTP) is used to promote misincorporation. An excess of Taq DNA polymerase is added to promote chain-extension beyond positions of base mismatch. Finally, serial dilutions between PCR cycles are performed to avoid predominance of a given mutation. Several independent mutagenic PCRs are performed in parallel and are then pooled prior to amplification by classical PCR, in order to maximize the variation of the introduced mutations.

The PCR fragments are subsequently subcloned into the appropriate two-hybrid vector, to create a mutant peptide library that can then be screened in a classical two-hybrid system, with the bait of interest. The complexity of the library to be obtained should be calculated as follows, for a 20 amino acid peptide encoded by 60 bp: for each base pair, there are three possibilities for introducing a mutation (for example, an A can be mutated into C, G, or T). This means that for one mutation/peptide approximately 180 clones (3×60) are required. Thus, in order to obtain a relevant number of approximately three mutations/insert, one needs a

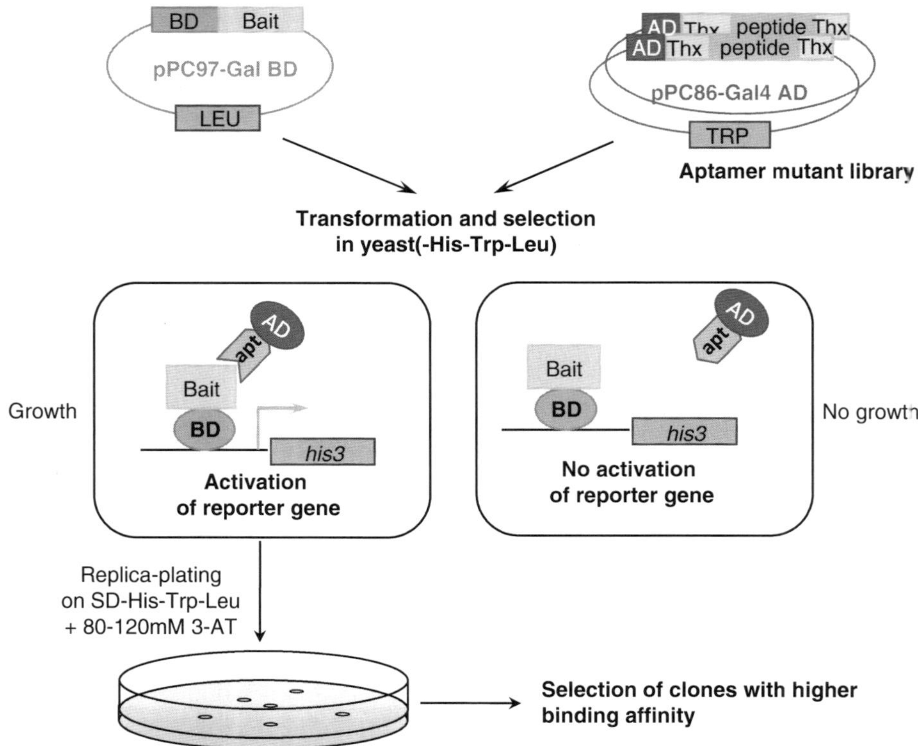

Fig. 2. Principle of the yeast two-hybrid screening using a strong selective drug.

library with a complexity of 5.8×10^6 and one should ideally screen approximately this number of clones to get a representative number of different mutations (keeping in mind that this includes also silent mutations and premature stop codons).

For the yeast-two hybrid screening, we have chosen a system in which the threshold of interaction detection can be modulated by the concentration of the 3-aminotriazole (3-AT) drug. This drug prevents Histidine synthesis, and thus only strong interactors, which will efficiently activate the Histidine reporter gene, will allow yeast to grow (Fig. 2). The concentration of 3-AT to be used varies for each aptamer/bait couple and has to be determined in a pilot yeast transformation assay prior to the actual large-scale mutant peptide library screening.

Following the mutant aptamer library screening, positive clones are picked on plates containing a concentration of 3-AT at which no interaction with the original peptide was detected anymore, and plasmid DNA from these clones is retransformed into yeast, in order to confirm strong binding to the bait protein under these stringent conditions. Sixty clones is a reasonable number to start with, considering possible redundancy and subsequent processing of clones.

Finally, the new aptamers obtained should be further processed in order to be tested in the same relevant biological assay as the initial aptamer from which they were derived. As an example, we present at the end of this chapter an in vitro GEF assay, in which we tested the optimized TRIP-like peptides we have obtained targeting the RhoGEF Tgat. We show that they are indeed at least two- to sixfold more efficient than the original peptide TRIPα in inhibiting Tgat exchange activity towards RhoA.

2. Materials

2.1. PCR-Based Random Mutagenesis

2.1.1. PCR-Based Mutagenesis

Solutions and reagents

1. Taq DNA Polymerase and 10× Taq polymerase Buffer.
2. 5′ and 3′ PCR primers: primers should be chosen annealing to flanking regions of the peptide coding sequence. Coding sequence overlapping base pairs will not be mutated.
3. PCR premix: mix 125 μl 10× Taq Pol buffer, 1.25 μl $MgCl_2$ 1 M, 2.5 μl dATP 100 mM, 2.5 μl dGTP 100 mM, 12.5 μl dCTP 100 mM, 12.5 μl dTTP 100 mM, 31.25 μl 5′ primer 20 μM, 31.25 μl 3′ primer 20 μM, qsp dH_2O 1 ml.
4. 3 M Sodium Acetate, pH 5.2.
5. 95 and 70% Ethanol.

2.1.2. PCR-Amplification of Mutated Inserts

Solutions and reagents

1. Pfu DNA polymerase and 10× Pfu polymerase Buffer.
2. 10 mM dNTP mix.
3. 5′ and 3′ PCR primers (same as in Subheading 2.1.1).
4. Appropriate restriction enzymes and their corresponding 10× reaction buffer.
5. Qiaquick Gel extraction Kit.

2.2. Construction of Mutant Aptamer Yeast Library

2.2.1. Ligation and Large-Scale Transformation into Competent *E. coli*

Strains and plasmids

1. Plasmid pPC86-Gal4AD-Thrx (Trp selection marker): we designed a modified version of the commercially available pPC86-Gal4AD plasmid in which we inserted the TrxA coding gene. Plasmid map can be obtained upon request.
2. *Escherichia coli* EP-Max 10B Electroporation F—mcrA Δ(mrr-hsdRMS-mcrBC) 80dlacZDM15 ΔlacX74 deoR recA1 endA1 araD139 Δ(ara, leu)7697 galU galK rpsL nupG.

Solutions and reagents

1. Appropriate restriction enzymes and their corresponding 10× buffers (New England Biolabs).
2. Alkaline Phosphatase (AP) 1 U/μl + 10× dephosphorylation buffer.
3. Phenol/chloroform.
4. Reagents for NaAC/ethanol precipitation (see Subheading 2.1.1).
5. Qiaquick Gel Extraction Kit.
6. T4 DNA Ligase (5 U/μl) and the appropriate 10× ligase buffer.
7. LB-medium: 10 g Bacto-Tryptone, 5 g Yeast extract, 5 g NaCl, qsp dH_2O 1 l. Autoclave.
8. Ampicillin: 100 mg/ml solution in dH_2O. This is a 1,000× stock solution.

Material

1. Electroporation cuvettes.
2. Electroporator *E. coli* Pulser.
3. 150 and 100 mm Petri dishes of LB agar + ampicillin.
4. Disposable cell scraper.

2.2.2. Large DNA Preparation and Sequencing of a Representative Number of Clones to Determine the Mutation Rate

Solutions and reagents

1. 50% Glycerol solution.
2. Qiagen Plasmid Maxi Kit.
3. Library efficiency DH5α chemically competent *E. coli* (10^8).
4. All reagents required for a plasmid extraction "miniprep" procedure (not described here).
5. Sequencing primers.

Material

1. 2 ml Cryotubes (NUNC).
2. 100 mm Petri dishes of LB agar + ampicillin.

2.3. Yeast Transformation and Yeast Two-Hybrid Screening Assay

2.3.1. Pilot Yeast Two-Hybrid Test in MAV103 Yeast Strain to Determine the Best Concentration of 3-AT

Strains and plasmids

1. Yeast strain: MAV103: *MAT***a** *SPAL10::URA3 leu2-3,112 trp1-901 his3D200 ura3-52 ade2-101 gal4Δ gal80Δ can1*^{r} *cyh2*^{r} *GAL1::HIS3@LYS2 GAL1::lacZ@URA3.*
2. Bait plasmid: pPC97-GalBD Bait (Leu selection marker).
3. Prey plasmid, i.e., original aptamer plasmid: pPC86-GalAD-Thrx-aptamer (Trp selection marker).

Solutions and reagents

1. YPD medium: 10 g Yeast extract, 20 g BactPeptone, 50 mg Adenine, qsp 1 l dH_2O. Autoclave the solution. Cool down and add 100 ml 20% Glucose.
2. 1× TE: 0.01 M Tris–HCl pH7.5, 1 mM EDTA. Autoclave the solution.
3. 1× LiAc: 0.1 M LiAc pH 7.5 (pH adjusted with glacial acetic acid). Autoclave solution.
4. Li-PEG: 40% PEG-4000 (stock solution: 50%), 0.1 M LiAc, 1× TE. Autoclave the solution.
5. Carrier DNA: salmon testes DNA 10 mg/ml, boiled and kept on ice.
6. SD medium: 2 g Yeast Nitrogen Base, 5 g NH4SO4, 20 g Agar. Qsp 800 ml dH_2O. Autoclave. Then add 100 ml of 20% glucose solution. Then add 20 ml of 50× Basic Supplement (1 g/l Arginine, 1.5 g/l Isoleucine, 1.5 g/l Methionine, 2.5 g/l Phenylalanine, 5 g/l Threonine). Then add 10 ml of each of the required marker amino acids depending on the selection required (see item 7).
7. Marker amino acid solutions (100×): 0.25% Histidine, 0.5% Leucine, 0.5% Lysine, 0.5% Tryptophane, 0.25% Uracile, 0.25% Adenine.
8. 1 M 3-AT (3-amino-1,2,4-triazole) solution.
9. SD-Leu-Trp medium: SD medium as above, in which 10 ml of each of the marker amino acid solutions are added, except Leucine and Tryptophane.
10. SD-Leu-Trp-His + 3-AT: SD medium as above, in which 10 ml of each of the marker amino acid solutions are added, except Leucine, Tryptophane, and Histidine. In addition, 3-AT is added at final concentration of 20–120 mM.

2.3.2. Large-Scale Transformation of Mutant Peptide Aptamer Library

Strains and plasmids

1. Yeast strain: MAV103: *MAT***a** *SPAL10::URA3 leu2-3,112 trp1-901 his3D200 ura3-52 ade2-101 gal4Δ gal80Δ can1*^{r} *cyh2*^{r} *GAL1::HIS3@LYS2 GAL1::lacZ@URA3.*
2. Bait plasmid: pPC97-GalBD Bait (Leu selection marker).
3. Library (see Subheading 2.2) inserted in pPC86-Gal4AD-Thrx plasmid (Trp selection marker). Reporter gene: His

Solutions and reagents

1. YPD medium.
2. SD-Leu selective medium.

3. All solutions and reagents for yeast transformation (see Subheading 2.3.1).
4. SD-Leu-Trp selective medium.
5. 1 M 3-AT (3-amino-1,2,4 triazole) solution.

Material

1. Velvets for replica plating.
2. 150 mm SD-Leu-Trp agar plates.
3. 150 mm SD-Leu-Trp-His + 80 mM (or 120 mM) 3-AT agar plates.

2.4. Isolation of High Affinity Clones and Inhibition Test In Vitro

2.4.1. Isolation of Positive Clones from Yeast

Solutions and reagents

1. SD-Trp selective medium.
2. Yeast extraction buffer (YEB): 10 mM Tris–HCl pH 8.0, 100 mM NaCl, 1 mM EDTA, 2% Triton-X100, 1% SDS.
3. Phenol:chloroform:isoamyl alcohol mix: 25:24:1 (Biosolve Ltd, The Netherlands).
4. Phenol/chloroform.
5. Acid washed glass beads.
6. 95% ethanol and 70% ethanol.
7. Library efficiency DH5α chemically competent *E. coli* (10^8).
8. LB medium + ampicillin (0.1 mg/ml).

Material

1. Wooden toothpicks, autoclaved.
2. LB agar plates + ampicillin (0.1 mg/ml).

2.4.2. Retransformation into Yeast to Confirm Interaction

Solutions and reagents

1. All solutions and reagents for LiAC yeast transformation (see Subheading 2.3.1).
2. SD-Leu-Trp agar plates.
3. SD-Leu-Trp-His + 80–120 mM 3-AT agar plates.

Material

1. Velvets for replica plating.

2.4.3. Sequencing of Clones

1. Clones are sent to Eurofins MWG Company for sequencing.

2.4.4. Processing of Clones for In Vitro Inhibition Assay

Subcloning of the selected aptamers into the pGEX vector
Solutions and reagents

1. pGEX vector (GE Healthcare).
2. Appropriate restriction enzymes and their corresponding 10× buffers.

3. Alkaline Phosphatase (AP) 1 U/μl + 10× dephosphorylation buffer.
4. Qiaquick Gel extraction Kit.
5. Reagents for ligation (see Subheading 2.2.1).
6. Sequencing primer for pGEX plasmid: pGEX 5′primer:

   5′GGGCTGGCAAGCCACGTTTGGTG-3′
   pGEX 3′primer: 5′-CCGGGAGCTGCATGTGTCAGAGG-3′
7. Library efficiency DH5α chemically competent *E. coli* (10^8).

Production of recombinant proteins.
Solutions and reagents

1. LB-medium + ampicillin (0.1 mg/ml).
2. IPTG (Isopropyl-β-d-thiogalactopyranoside; Euromedex): 1 M solution in dH_2O.
3. Glutathione Sepharose beads.
4. Lysis buffer: 50 mM Tris pH 7.5, 50 mM NaCl, 5 mM $MgCl_2$, 1 mM DTT, 1 mM AEBSF, 1 mg/ml leupeptin/aprotinin. For GST-GTPase production, add 10 μM GDP to this buffer.
5. Elution Buffer: 20 mM Tris–HCl pH 8.0, reduced Glutathione 10 mM. Buffer to pH 8.0 with NaOH. For GST-GTPase production, add 10 μM GDP, 2 mM $MgCl_2$ to this buffer.
6. 10% SDS-PAGE gel and required running buffer.
7. 50% Glycerol solution.

Material

1. Round centrifuge tubes (50 ml capacity).
2. Sonicator.
3. Standard SDS PAGE gel electrophoresis material.

2.4.5. In Vitro Guanine Nucleotide Exchange Assay

Solutions and reagents

1. Loading buffer: 20 mM Tris–HCl pH 7.5, 50 mM NaCl, 2 mM $MgCl_2$, 1 mM DTT, 50 μg/ml BSA, 1 μM mant-GTP (*N*-methylanthraniloyl fluorophore; Molecular Probes).

Material

1. 96-Well plates.
2. FLX800 Microplate Fluorescence Reader.

3. Methods

3.1. PCR-Based Random Mutagenesis

3.1.1. PCR-Based Mutagenesis

The mutagenic PCR is performed starting with 1 ng of template plasmid DNA of the aptamer of interest and using specific conditions, in order to maximize misincorporation of base pairs (see Subheading 1). The detailed protocol is described in Fig. 3.

- 7 mM $MgCl_2$ is used to stabilize noncomplementary base pairs.
- 640 μM $MnCl_2$ is added to diminish the template specificity of the Taq DNA polymerase.
- 1 mM dCTP and dTTP are used to promote misincorporation, while dATP and dGTP are kept at 0.2 mM.

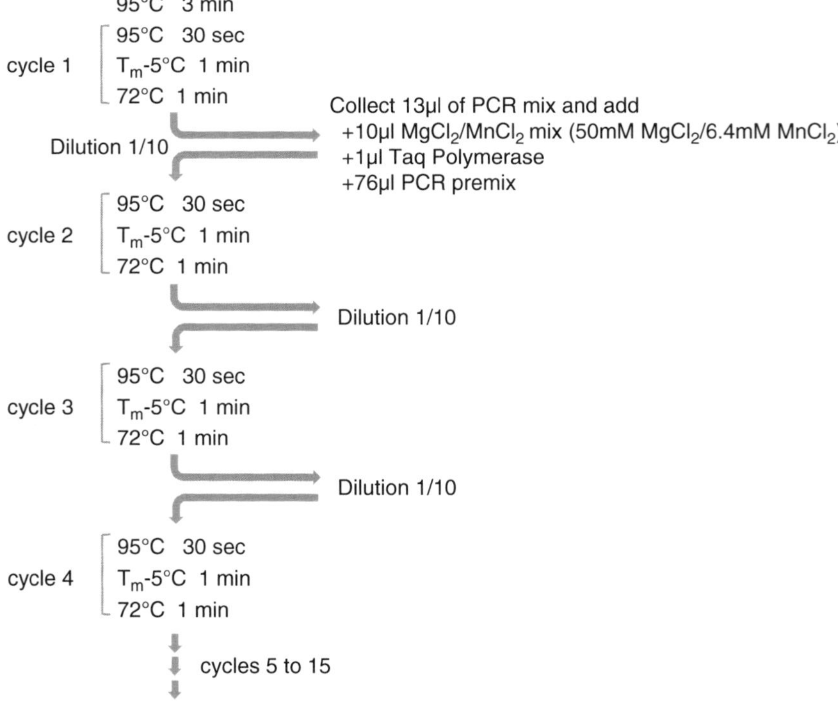

Fig. 3. PCR-based random mutagenesis protocol.

- Two times excess (5 U/reaction) Taq DNA polymerase is added to promote chain-extension beyond positions of base mismatch.
- Finally, serial dilutions between PCR cycles are performed to avoid predominance of a given mutation.

In order to maximize the variation of the introduced mutations, at least six independent mutagenesis PCRs are performed in parallel and are then pooled prior to the next step (PCR-amplification, Subheading 3.1.2) (see Note 1). Using these PCR conditions, we have obtained a mean mutation rate of 3.2 mutations/peptide (range from 1 to 6), which is sufficient to obtain a satisfactory panel of diverse peptides (as determined by sequencing a representative number of fragments) (see Note 2).

To minimize the final volume of the pooled PCR products, it is recommended to ethanol precipitate them prior to amplification.

1. Add of 1/10 volume of 3 M sodium acetate pH 5.2 and 2.5 volumes of ice cold 100% ethanol.
2. Mix well and place at −20°C for >20 min.
3. Spin at maximum speed in a microfuge for 10–15 min.
4. Carefully decant the supernatant and add 1 ml 70% ethanol.
5. Mix well and spin briefly. Carefully decant the supernatant.
6. Air dry or briefly vacuum the dry pellet. Resuspend the pellet in 50 μl water.

3.1.2. PCR-Amplification of Mutated Inserts

Following this PCR, which should yield up to 300 ng of pooled mutagenized peptide DNA (in 50 μl final volume), a classical PCR is performed, in order to amplify the mutated inserts. Five PCRs are performed in parallel, each done with 10 μl PCR product.

1. Mix 10 μl of mutagenized PCR product with 5 μl 10× Pfu polymerase buffer, 1 μl dNTP mix (10 mM), 5 μl 5′primer 2 μM, 5 μl 3′primer 2 μM, 1 μl Pfu. Complete to 50 μl with dH_2O.
2. PCR cycles: 95°C 5 min, 95°C 30 s, 51°C 1 min, 72°C 1 min, (40 cycles), 72°C 10 min.
3. Check an aliquot of the PCR on a 2% agarose gel, in order to verify efficient amplification of the mutagenized DNA.

The PCR products are then pooled again in an eppendorf tube, and digested with the appropriate restriction enzymes, in order to create cohesive ends, for subsequent cloning into the prey plasmid pPC86-Gal4AD-Thrx (see Subheading 3.2.1 for vector preparation).

1. To 250 μl PCR products, add 40 μl 10× enzyme buffer + 5 μl of each enzyme, qsp 400 μl dH_2O.

2. Incubate 2 h at 37°C, then load all on a 2% agarose gel, and purify DNA fragment-containing agarose bit using Qiaquick Gel extraction Kit according to manufacturer's instructions (Qiagen). Elute from column twice with 40 μl H_2O.
3. Load 1 μl of purified insert on a 2% agarose gel, to check the size and concentration of the insert.

3.2. Construction of Mutant Aptamer Yeast Library

3.2.1. Ligation and Large-Scale Transformation into Competent *E. coli*

pPC86-Gal4AD-Thrx vector preparation

1. Cut 20 μg vector in 300 μl final volume, with 30 μl 10× enzyme buffer + 5 μl of each enzyme.
2. Incubate 2 h at 37°C, then load all on a 2% agarose gel and purify vector containing agarose fragment using Qiaquick Gel extraction Kit according to manufacturer's instructions. Perform two successive elutions from the same column (2 × 50 μl).
3. Dephosphorylate vector ends using alkaline phosphatase. Add 10 μl 10× AP buffer directly into the eluted vector and add 2 μl AP.
4. Incubate 20 min at 37°C, then add 2 μl AP again and incubate 10 more minutes.
5. Heat inactivate the AP by adding 20 mM EGTA (final concentration) and incubating for 10 min at 65°C.
6. Phenol/chloroform extract vector DNA by adding 1 volume phenol/chloroform (see Note 3), vortexing and spinning for 5 min at full speed in a microfuge.
7. Transfer the upper aqueous phase into a new tube and add 1 volume of chloroform. Vortex and spin again as above.
8. Ethanol precipitate the supernatant by proceeding as above (Subheading 3.1.1, steps 1–6).
9. Purify again this DNA suspension on a Qiaquick column, and elute in 50 μl H_2O.
10. Check 1 μl of purified vector on a 1% agarose gel, to check size and concentration of vector (see Note 4). For best results, aim at obtaining at least a concentration of 200 ng/μl of vector.

Ligation. The ligation is performed as many independent ligation reactions, to obtain the maximum possible representation of clones. Reactions are then pooled to transform competent *E. coli*. The vector/insert molar ratio should be 1:3 (or up to 1:5).

1. Prepare 40 eppendorf tubes, each one containing the following mix: 200 ng vector, 10–20 ng insert (purified PCR products of Subheading 3.1.2), 1 μl 10× ligation buffer, 1 μl high conc ligase, qsp 10 μl H_2O (see Note 5).
2. Prepare one additional tube (negative control) containing the same mix as above except insert DNA.

3. Incubate these ligation mixes at 12°C, overnight in a water bath (see Note 6).

Transformation of ligation mixes into ultra competent *E. coli*.

1. Before transforming into bacteria, the ligation reactions should be pooled and precipitated with ethanol, using the protocol described in Subheading 3.1.1 (steps 1–6). Resuspend in 200 μl final volume H_2O (see Note 7).
2. Transform ultra electro-competent *E. coli* (EP Max 10^{10}), which have revealed the most efficient in our hands for this purpose. Transform 40 × 5μl of the precipitated ligation mix (+5 μl of the control ligation) into 41 × 50 μl bacteria, according to the manufacturer's instructions.
3. For plating of the transformed bacteria, prepare 20 large 150 mm plates of LB agar + ampicillin (0.1 mg/ml). Plate two tubes on the same plate.
4. In parallel, plate 1/500 and 1/1,000 of a transformation on two 100 mm LB agar + ampicillin plates (+ the negative control). These plates will serve for counting of the number of transformants, i.e., for determining the number of clones contained in the library (see Note 8).
5. Let grow overnight at 37°C.
6. The next day, a lawn of bacteria should have grown on each plate, and no colony should have appeared on the control plate. Using a cell scraper, scrape off this lawn from each plate in 5–10 ml LB and collect the bacterial suspension in a sterile flask. This yields approximately 250 ml of suspension. This can further be diluted with fresh LB + ampicillin for better growth results, up to 500 ml final volume.
7. Let this culture grow over night in a shaking incubator at 37°C.

3.2.2. Large DNA Preparation and Sequencing of a Representative Number of Clones to Determine the Mutation Rate

1. Before proceeding to the plasmid DNA extraction, it is advisory to remove a 100 ml aliquot of the above culture and to freeze down glycerol stocks of the library in cryotubes (in 20% glycerol). Store the tubes (approximately 100) at –80°C.
2. Split the rest of the culture into 4× 100 ml and proceed with the DNA extraction using four columns of the Qiagen MAXIprep Kit, according to manufacturer's instructions. In total, this yields at least 1 mg library plasmid DNA, at 1 μg/μl.
3. At this stage, it is crucial to verify the successful creation of the library. To do so, 1 μg of library plasmid DNA is transformed into classical DH5α bacteria, plated on LB agar + ampicillin, and 24 DNA minipreps are prepared from picked single colonies. These are sent to sequencing.

3.3. Yeast Transformation and Yeast Two-Hybrid Screening Assay

The optimization strategy described in this chapter relies on the rationale that stronger binding of the peptides to the bait in yeast will also lead to stronger inhibition. Therefore the yeast screening is performed on selective medium (for both bait and prey, i.e., SD-Leu-Trp) to which the drug 3-AT is added, which prevents Histidine synthesis. Thus only strong interactors, which strongly activate the Histidine reporter gene, will allow yeast to grow. The concentration of 3-AT to be used varies for each peptide/bait couple and has to be determined in a yeast transformation assay prior to the actual large-scale mutant peptide library screening. A concentration should be chosen at which no interaction with the original aptamer is detected anymore.

3.3.1. Pilot Yeast-Two Hybrid Test in MAV103 Yeast Strain to Determine the Best Concentration of 3-AT Drug

Classical yeast transformation

1. Day 0: set up a 10 ml culture of MAV103 yeast in YPD medium in a 50 ml Falcon tube, and let grow overnight in a 30°C shaking incubator.
2. Day 1: dilute the starter to 1/20 in the required volume of YPD (see step 5) and let grow to early log phase, i.e., OD_{600nm} of 0.4–0.5, which corresponds to $5–7 \times 10^6$ cells/ml.
3. Spin down yeasts in 50 ml Falcon tubes, 4 min at 2,000×*g*, at room temperature.
4. Wash the pellet once with sterile dH_2O and pool everything in one tube.
5. Spin again as above and resuspend the pellet in 0.1 M LiAc-TE, at a concentration of 5×10^8 cells/ml. The volume depends on the number of transformations that have to be done. 100 μl of this yeast suspension is required per transformation. So when setting up the yeast culture, plan enough volume to have enough of the final suspension at 5×10^8 cells/ml.
6. Prepare as many 1.5 ml eppendorf tubes as required, containing 2 μg of the DNA to be transformed (up to 10 μl volume) and 4 μl of carrier DNA. Classical transformation conditions include a negative control (no DNA), the bait plasmid alone, the prey peptide plasmid alone, and bait/peptide plasmids together.
7. Add 100 μl per tube of the yeast suspension of step 5.
8. Add 600 μl Li-PEG solution and mix by flicking the tube several times until a homogenous suspension is obtained (see Note 9).
9. Incubate the tubes for 30 min at 30°C. In the meantime preheat a water bath to 42°C.
10. Perform the heat-shock for 10 min at 42°C.
11. Spin yeast briefly in a microfuge (5 s, full speed).
12. Wash the pellet with 1× TE by pipetting up and down and spin again as above.

13. Finally, resuspend the pellet in 150 μl 1× TE and spread all on SD-Leu-Trp selective plates. Plates are then incubated for 3 days in a 30°C incubator.

Replica plating of grown colonies on selective medium containing increasing concentrations of 3-AT.

1. Prepare SD-Leu-Trp-His agar plates, containing increasing concentration of 3-AT (see Note 10). Starting from a 1 M stock solution of 3-AT, dilute accordingly to obtain plates containing 0, 20, 40, 60, 80, 100, 120 mM 3-AT.
2. Replica plate the colonies grown on SD-Leu-Trp plates on each of these plates and incubate at 30°C, until colonies appear (3–4 days).
3. Chose the concentration of 3-AT to be used for the large-scale screening by determining the plates on which no colonies grow, i.e., at which no interaction with the original aptamer is detected anymore (see Note 11).

3.3.2. Large-Scale Transformation of Mutant Peptide Aptamer Library

For the large-scale yeast transformation with the created mutant peptide library, transformation efficiency is critical, in order to obtain a maximum number of transformants. Therefore it is recommended to transform yeast sequentially, i.e., first with the bait plasmid alone, and subsequently with the library. This has been shown to significantly increase transformation efficiency.

1. Transform MAV103 yeast with pPC97-GalBD Bait plasmid using the protocol detailed in Subheading 3.1.1, and plate on SD-Leu selective medium. Plates are then incubated for 3 days in a 30°C incubator.
2. Pick one colony from this plate (MAV103-pPC97-Bait) and inoculate a 100 ml culture in SD-Leu medium, as a starting culture for the large-scale transformation.
3. Large-scale yeast transformation: this transformation basically consists in a many fold of single transformations, using the classical yeast transformation protocol described in Subheading 3.1.1, with the slight modifications described below.
4. Dilute the starter to 1/20 in 1,000 ml of SD-Leu selective medium in an appropriate flask and let grow to early log phase, i.e., OD_{600nm} of 0.4–0.5, which corresponds to $1–2 \times 10^6$ cells/ml (in synthetic medium).
5. Centrifuge and wash the yeast cells as above and resuspend them in 4.2 ml 0.1 M LiAc-TE. This gives a cell suspension of 5×10^8 cells/ml.
6. Prepare 42 eppendorf tubes:

40 tubes: each containing 10 μl of the mutant peptide library (in the pPC86-Gal4AD-Thrx vector) (see Subheading 3.2) and 4 μl of carrier DNA.

One tube containing no DNA and one tube containing the original pPC86—wt aptamer plasmid and 4 μl of carrier DNA.

7. For the rest of the protocol, proceed as in Subheading 3.3.1. (steps 7–13), except that after resuspending the yeast pellet in 150 μl 1× TE, eight tubes are pooled together and plated on 150 mm SD-Leu-Trp selective plates (5 plates total) (see Note 12). The control transformations are processed as above, and plated on 100 mm plates of the same medium.

   Using this procedure, we screened approximately 3×10^5 clones.

8. Five days after plating, yeasts are replica plated on SD-Leu-Trp-His selective plates, containing the concentration of 3-AT determined in the pilot yeast assay (in our case 80 and 120 mM).

3.4. Isolation of High Affinity Clones and Inhibition Test In Vitro

3.4.1. Isolation of Positive Clones from Yeast

1. Pick colonies that grew on the plates containing 80 mM and 120 mM 3-AT, respectively.
2. Grow small cultures of each picked colony overnight in a 30°C shaking incubator, in 2 ml medium that maintains selection only for the prey plasmid DNA (SD-Trp) (see Note 13).
3. Fill a 1.5 ml eppendorf tube with the culture and collect the cells by a 5-s centrifugation in a microfuge.
4. Discard supernatant and briefly vortex the tube to resuspend the pellet in the residual liquid.
5. Add 200 μl of yeast extraction buffer (YEB). Add 200 μl phenol:chloroform:isoamyl alcohol (25:24:1) (see Note 3). Add 200 μl of acid-washed-beads.
6. Vortex for 5–7 min and centrifuge for 5 min in a microfuge.
7. Collect aqueous phase and perform second phenol/chloroform extraction.
8. Vortex briefly and centrifuge again.
9. Remove 100 μl of aqueous phase, transfer into new tube and precipitate it with Ethanol (see Note 14). Cool for 10 min at −20°C.
10. Centrifuge for 10 min in a microfuge, discard supernatant, wash the pellet with 200 μl 70% Ethanol, and centrifuge again for 1 min.
11. Discard supernatant and let pellet briefly air-dry.

12. Resuspend the pellet in 50 μl dH_2O and use 1 μl to transform chemically competent *E. coli*, according to manufacturer's instructions (see Note 15), plating on LB + ampicillin.
13. Plasmid DNA is extracted from grown colonies, following standard "miniprep" procedures.

3.4.2. Retransformation into Yeast to Confirm Interaction

Using the Li-Ac yeast transformation protocol (Subheading 3.1.1), it is important to confirm that the selected and isolated prey plasmids interact with the bait on the stringent 3-AT medium. Proceed as described in Subheading 3.1.2 for replica plating.

3.4.3. Sequencing of Clones

Positive clones, whose strong binding to bait protein under stringent conditions has been confirmed upon retransformation into yeast, are then sequenced to determine the mutated codon and deduced amino acid sequence. 60 clones is a reasonable number to start with, considering possible redundancy and subsequent processing of clones. Two micrograms plasmid DNA from a miniprep are sent to sequencing, together with the appropriate 5′ primer.

3.4.4. Processing of Clones for In Vitro Inhibition Assay

After sequencing and chosing the clones of interest, aptamers should be further processed in order to test them in the relevant biological assay. In the example we present here, aptamers are produced as recombinant proteins and subsequently tested in an in vitro guanine nucleotide exchange assay, with recombinant RhoA GTPase and Tgat GEF.

Subcloning of the selected aptamers into the pGEX vector.

Aptamer inserts are subcloned into the pGEX-5X bacterial expression vector, for further recombinant protein production, using classical subcloning procedures. Briefly:

1. In an eppendorf tube, digest plasmid DNA pPC86-Aptamer-X with appropriate restriction enzymes to cut out insert: 20 μl DNA (of a miniprep) + 5 μl 10× enzyme buffer + 1 μl of each enzyme, qsp 50 μl dH_2O.
2. In parallel, digest pGEX-5X1 vector with appropriate restriction enzymes to linearize vector: 5 μg DNA + 5 μl 10× enzyme buffer + 1 μl of each enzyme, qsp 50 μl dH_2O.
3. Incubate 1–2 h at 37°C and verify 5 μl on a 1% agarose gel, to check for correct linearization of the vector and correct size of the insert.
4. Dephosphorylate vector ends using alkaline phosphatase. Add 6 μl 10× AP buffer directly into the digestion reaction + 1.5 μl AP.
5. Incubate 20 min at 37°C, then add 0.5 μl AP again, and incubate 10 more minutes.
6. Load both vector and insert on an agarose gel, and purify DNA containing agarose fragment using the Qiaquick Gel extraction

Kit according to the manufacturer's instructions. Elute the insert in 30 µl H_2O and the vector in 50 µl.

7. Check 1 µl of the purified insert and vector on an agarose gel, to check the size and concentration.
8. Purified vector and inserts are then ligated together, at a molar ratio of 1:3 (estimation on agarose gel is sufficient). Final volume of reaction: 10 µl. Mix all reagents on ice.
9. Incubate ligation reaction at 12°C in a water bath for at least 4 h, better overnight.
10. Ligation reactions are then transformed into chemically competent *E. coli* DH5α, according to manufacturer's instructions, and plated on LB + ampicillin plates.
11. Plasmid DNA extraction "minipreps" are performed on colonies picked from the plates.
12. Positive clones are then sent to sequencing with the pGEX5′ primer, to verify the correct in frame insertion of the insert.

Production of recombinant proteins. GST-GEFs and GST-GTPases are produced using the standard protocol described below, except that GTPases are produced and eluted in the constant presence of 10 µM GDP and 2 mM $MgCl_2$ in the buffer (= GDP-loaded GTPase).

1. Set up 50 ml preculture in LB + ampicillin overnight at 37°C, in a shaking incubator.
2. Dilute the starter in 1 l LB + ampicillin, in a large flask. Grow at 37°C in a shaking incubator until OD_{600nm} of 0.8 (up to 1) is reached. Add IPTG to 0.1 mM final concentration to induce protein production and incubate 3 h at 20°C, in a shaking incubator.
3. Spin at 4,000×*g* for 15 min, 4°C (in big buckets).
4. Resuspend the pellet in 10 ml 1× PBS and transfer into 50 ml Falcon tubes and spin again 2,000×*g* for 15 min at 4°C (see Note 16).
5. Resuspend the pellet in 10 ml ice cold lysis buffer, transfer into 50 ml tubes (Nalgene).
6. Sonicate 3× 30 s (40% intensity).
7. Centifuge for 10 min at 20,000×*g* at 4°C.
8. Transfer the supernatant in a 50 ml Falcon tube, on ice.
9. Resuspend the remaining pellet again in 10 ml lysis buffer. Sonicate and centrifuge as above.
10. Collect the supernatant and pool with the first supernatant in a Falcon tube, on ice.
11. Add 350 µl Glutathione Sepharose beads (50% slurry equilibrated in lysis buffer).

12. Leave on wheel for 30 min at 4°C and then centrifuge for 3 min at 2,000×*g*, at 4°C.
13. Wash pellet once with 5–10 ml cold lysis buffer (without protease inhibitors and DTT).
14. Resuspend the beads in 1 ml lysis buffer and transfer into an eppendorf tube.
15. Wash twice with 1 ml lysis buffer, by centrifuging at 16,000×*g* for 20 s.
16. After the last spin, elute the bound protein by adding 400 μl Elution buffer and incubating on a rotating wheel for 1 h at 4°C.
17. Centrifuge as above and transfer supernatant into an eppendorf tube.
18. Check all proteins on SDS-PAGE gel together with a BSA standard, to verify correct size and concentration of the produced proteins.
19. Add glycerol to 30% final concentration and snap freeze the proteins in aliquots in liquid N_2, store at –80°C (see Note 17).

3.4.5. In Vitro Guanine Nucleotide Exchange Assay

In order to evaluate the inhibitory potential towards their GEF target of the newly isolated, stronger interacting aptamers, a fluorescent guanine nucleotide exchange assay is performed, using the recombinant GST-fusion proteins produced in Subheading 3.4.4. Nucleotide exchange is monitored by measuring mant-GTP loading on the GTPase.

1. Premix: pre-incubate GEF and aptamer in 95 μl loading buffer containing mant-GTP. The GEF (0.1–0.5 μM) is allowed to equilibrate with 20 μM peptide inhibitor for 5 min in loading buffer containing 1 μM mant-GTP, in a 96-well plate, at 25°C.
2. Start the reaction by adding 5 μl of 20 μM GDP-loaded GST-GTPase to the premix, in order to obtain a final concentration of 1 μM GTPase, in 100 μl final volume.
3. The exchange reaction is monitored for 15 min (measurements are taken every 10 s) by following the increase in relative fluorescence due to the loading of the fluorescent GTP analog upon binding to the GTPase at 25°C, using a FL_X800 microplate fluorescence reader (BIO-TEK Instruments) (λ_{ex} = 360 nm, λ_{em} = 460 nm).

An example of the TRIP-derived peptides we obtained that showed stronger interaction and stronger inhibition potential is shown in Fig. 4.

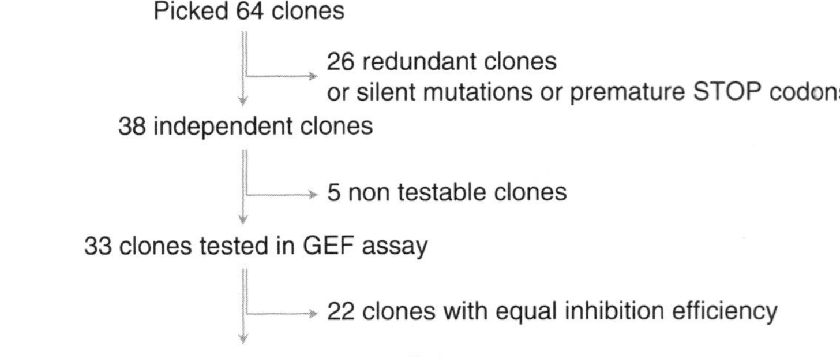

PEPTIDIC SEQUENCE	MUTANT	[3-AT]	Fold over TRIP Inhibition
AREGADGAICGYNLATLVMLGPSERVFCPLCEPCSSDIYELM	**WT**	30mM	1
AREGADGAICGYNLAMSVMLGPSERVFCPLCEPCSSDIYELM	**1**	120mM	5,5 X
AREGADGAICGYNLATLDMLGPSERVLCPLCEPCSSDIYELM	**2**	120mM	3,4 X
AREGADGAICGYNLATLVMLGPSERVFCPLCGPCSTDIYELM	**3**	120mM	3,1 X
AREGADGAICGYNLAALGMLGPSERVFCPLCGPCSSDNYELM	**4**	120mM	2,5 X
AREGADGAICGYDLAMLVMLGPSERVFCPLCEPRSSDIYELM	**5**	120mM	1,5 X
AREGADGAICGYNLATLVMLGPSERVFCPLCGPCSSDIYELM	**6**	80mM	6 X
AREGADGAICGYNLAMLVMLGPGERVFCPLCEPCSSDIYELM	**7**	80mM	3,5 X
AREGADGAICGYNLATLAMLGPSARVFCPLCGPCSSDIYELM	**8**	80mM	2,3 X
AREGADGAICGYNLAMLVMLGPSERVFCPLCEPCSSDIYELM	**9**	80mM	2 X
AREGADGAICGYNLAASVMLGPSERVFCPLCEPRSPDIYELM	**10**	80mM	1,4 X
AREGADGAICGYNLATLDMLGPSERVFCPLCEPCSSDSYELM	**11**	80mM	1,1 X

Fig. 4. Example of TRIP-derived optimized peptide aptamers obtained with the method described in this chapter.

4. Notes

1. It is preferable to split the PCR master mix into different aliquots, to run these reactions separately and then to mix them together upon completion of the reaction (before the amplification PCR reaction), in order to prevent overrepresentation in the library of single mutations occurring in the first rounds of PCR. This will increase the diversity of the library.
2. In order to determine whether a satisfactory mutation rate has been obtained, it is advised to subclone an aliquot of the PCR fragments into the TOPO cloning vector (TOPO TA Cloning Kit for Sequencing, Invitrogen) according to the manufacturer's instructions and to sequence a relevant number of clones, before constructing the library.
3. Phenol is highly corrosive and can cause severe burns. Chloroform is irritating to the skin, eyes, mucus membranes, and upper respiratory tract. It is carcinogen and may damage liver and kidneys. Therefore, gloves, protective clothing, and safety glasses should be worn when handling them, and work in a chemical hood is strongly advised.

4. Experience has shown that in order to precisely determine DNA concentration of vector (and insert), measurement on a device such as the Nanodrop is less precise than an evaluation on agarose gel, using a standard DNA as control.
5. In some cases, we have noticed that adding 5% PEG 4000 to the ligation mix enhances ligation efficiency.
6. Incubation of the ligations at 12°C has proven more efficient than at 16 or 4°C or even RT.
7. Precipitation of the ligations in order to get rid of any salts contained in the reaction is crucial to avoid the formation of an electric arc when electroporating the bacteria.
8. Typically, using this protocol, we obtained a library complexity of at least 1×10^6.
9. Do not vortex the tubes, as PEG renders yeast fragile.
10. Caution: 3-AT is a highly toxic compound and should be handled accordingly, with all protective and safety measures.
11. In our case, 40 mM 3-AT was the highest concentration at which colony growth was observed with the initial "wild type" aptamer. Therefore we chose concentrations of 80 and 120 mM 3-AT for the large-scale screening.
12. It is recommended to plate an aliquot of the final yeast suspension (1/10 of a 150 μl suspension) on a separate 100 mm selective plate. This will allow determining the transformation efficiency, by counting the number of colonies grown on this plate. In our case, we screened approximately 3×10^5 clones.
13. When growing yeast for plasmid DNA extraction, grow the yeast in the medium that selects only for the prey plasmid (i.e., SD-Trp), don't keep the selection of the bait plasmid (Leu). This has turned out more efficient to extract prey plasmid (over bait plasmid).
14. Be careful when collecting the aqueous phase from the tube, to make sure no phenol/chloroform is contaminating the collected phase.
15. Usually, transformation into *E. coli* of plasmid DNA isolated from yeast gives low yields. To improve the transformation efficiency, one can use up to 5 μl of isolated plasmid, and it is strongly recommended to use highly competent, commercially available bacteria (competence of at least 10^8).
16. This is a good moment to freeze the bacterial pellets at −20°C to do the extracts later.
17. Stability of recombinant proteins stored at −80°C is highly variable and should be determined for each type of protein. Usually, they are stable for at least 1 month, and up to 1 year.

Acknowledgments

We are grateful to Anne Briançon-Marjollet for her contribution to the library construction. This work was supported by grants from the ANR "Physique et chimie du vivant" grant N° 06-137373. N. B. was supported by a CNRS valorization fellowship from CNRS. All authors are members of the CNRS consortium GDR2823.

References

1. Colas P, Cohen B, Jessen T, Grishina I, McCoy J, Brent R (1996) Genetic selection of peptide aptamers that recognize and inhibit cyclin-dependent kinase 2. Nature 380:548–50
2. Crawford M, Woodman R, Ko Ferrigno P (2003) Peptide aptamers: tools for biology and drug discovery. Brief Funct Genomic Proteomic 2:72–9
3. Hoppe-Seyler F, Crnkovic-Mertens I, Tomai E, Butz K (2004) Peptide aptamers: specific inhibitors of protein function. Curr Mol Med 4:529–38
4. Borghouts C, Kunz C, Groner B (2005) Peptide aptamers: recent developments for cancer therapy. Expert Opin Biol Ther 5:783–97
5. Schmidt S, Diriong S, Mery J, Fabbrizio E, Debant A (2002) Identification of the first Rho-GEF inhibitor, TRIPalpha, which targets the RhoA-specific GEF domain of Trio. FEBS Lett 523:35–42
6. Bouquier N, Fromont S, Zeeh JC, Auziol C, Larrousse P, Robert B, Zeghouf M, Cherfils J, Debant A, Schmidt S (2009) Aptamer-derived peptides as potent inhibitors of the oncogenic RhoGEF Tgat. Chem Biol 16:391–400
7. Yoshizuka N, Moriuchi R, Mori T, Yamada K, Hasegawa S, Maeda T, Shimada T, Yamada Y, Kamihira S, Tomonaga M, Katamine S (2004) An alternative transcript derived from the trio locus encodes a guanosine nucleotide exchange factor with mouse cell-transforming potential. J Biol Chem 279:43998–4004
8. Cadwell RC, Joyce GF (1992) Randomization of genes by PCR mutagenesis. PCR Methods Appl 2:28–33

Chapter 9

A Screening Strategy for Trapping the Inactive Conformer of a Dimeric Enzyme with a Small Molecule Inhibitor

Charles S. Craik and Tina Shahian

Abstract

Kaposi's sarcoma-associated herpesvirus (KSHV) is the etiological agent of Kaposi's sarcoma (KS), the most common cancer in AIDS patients. All herpesviruses express a conserved dimeric serine protease that is required for generating infectious virions and is therefore of pharmaceutical interest. Given the past challenges of developing drug-like active-site inhibitors to this class of proteases, small-molecules targeting allosteric sites are of great value. In light of evidence supporting a strong structural linkage between the dimer interface and the protease active site, we have focused our efforts on the dimer interface for identifying dimer disrupting inhibitors. Here, we describe a high throughput screening approach for identifying small molecule dimerization inhibitors of KSHV protease. The helical mimetic, small molecule library used, as well as general strategies for selecting compound libraries for this application will also be discussed. This methodology can be applicable to other systems where an alpha helical moiety plays a dominant role at the interaction site of interest, and in vitro assays to monitor function are in place.

Key words: Human herpesvirus protease, Kaposi's sarcoma-associated herpesvirus (KSHV), Dimer disruptor, Allosteric inhibitor, Fluorogenic enzyme assay, Alpha helical mimetic molecules, High throughput screening

1. Introduction

Herpesviruses make up one of the most prevalent viral families including eight human types that cause a variety of devastating illnesses. The standard course of treatment for common herpesviral infections is a class of broad-acting viral DNA replication inhibitors, which exhibit undesirable toxicity, poor oral bioavailability, and in some cases inadequate efficacy. Efforts by pharmaceutical companies to target the active site of the essential dimeric serine protease of human herpesviruses (HHV) have yet to yield a drug-like candidate (1–6). Given the evidence supporting a conformational linkage

Yi Zheng (ed.), *Rational Drug Design: Methods and Protocols*, Methods in Molecular Biology, vol. 928,
DOI 10.1007/978-1-62703-008-3_9,

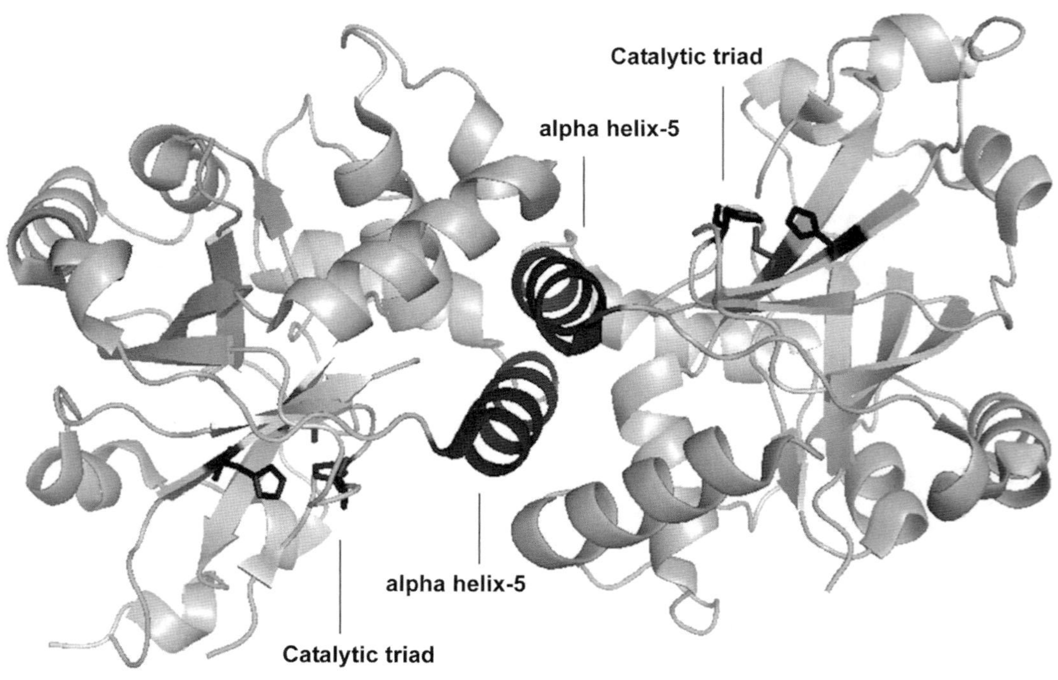

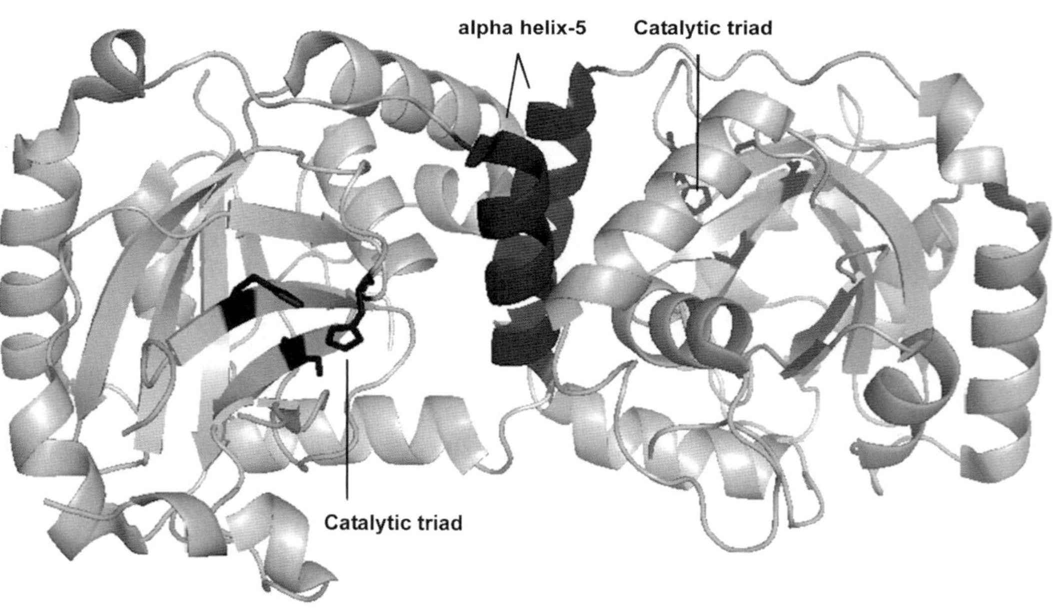

Fig. 1. KSHV Pr dimer. The *top and side views* of the dimeric crystal structure are shown. The interfacial α-helix 5 moieties and the two independent catalytic triads are highlighted in *black*.

between protease dimerization and activation, we have focused our efforts on identifying molecules that target the dimer interface (7–12). In the case of KSHV protease (KSHV Pr) the dimer interface covers approximately 2,500 Å^2 and includes the α-helix 5 of each monomer as the major constituent (Fig. 1) (13). In vitro

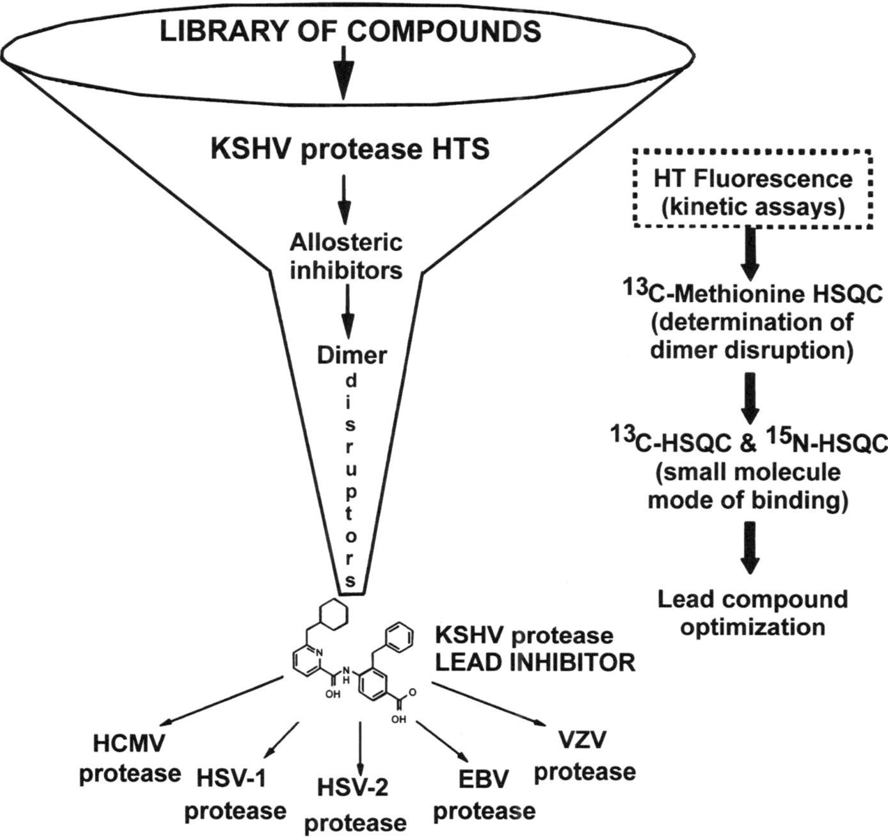

Fig. 2. The workflow for identifying dimer disruptors of HHV proteases. Inhibitors of KSHV Pr activity are identified by high throughput screening (HTS) of small molecule libraries. Hits are further characterized for mode of inhibition and dimer disruption through 2D-NMR assays that have been described previously. Lead compounds are then tested for efficacy against other HHV proteases. This method will focus on the HTS assay (*dotted box*).

studies with KSHV Pr and other HHV proteases have shown the dimer interface to be very sensitive to genetic perturbation, where single point mutations often lead to a loss of dimerization and activity (11). Furthermore the dimerization affinity is weak, with a reported K_D of 1.7 μM for KSHV Pr (10). These characteristics define the dimer interface as a suitable candidate for dimerization inhibitors. We first tested this approach by inserting the key interfacial α-helix 5 residues on the internally stabilized α-helix of a mini-protein (14). The resulting macromolecule disrupted the KSHV Pr dimer and inhibited enzyme activity, proving that targeting the dimer interface is a viable route for identifying novel inhibitors. In order to identify small molecule dimer disruptors of KSHV Pr, a workflow was developed that begins with high throughput screening (focus of this chapter), and continues with experiments that assess dimerization, mode of binding, and broad specificity against other HHV proteases (Fig. 2) (15, 16). Since two α-helices are the major component of the KSHV Pr dimer interface, we screened a library of helical mimetic small molecules. This library

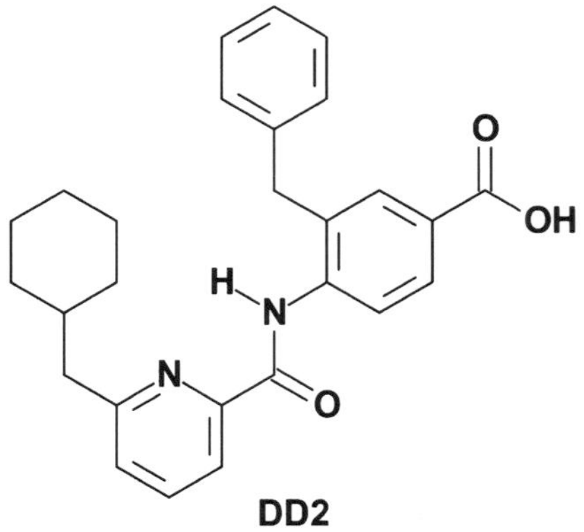

Fig. 3. The structure of DD2.

was composed of roughly 200 compounds that were originally developed by computational design to disrupt the interacting α-helix of p53 tumor suppressor protein with oncoprotein MDM2 (17). Screening was performed in a 96-well plate format using a fluorogenic activity assay. The substrate used is an optimized hexapeptide attached to 7-amino-4-carbamoylmethyl coumarin (ACC), where cleavage at the scissile bond releases the ACC group resulting in increased fluorescence (7, 12). We have successfully used this approach to identify the small molecule inhibitor DD2 (Fig. 3), which binds a novel allosteric pocket at the dimer interface of KSHV Pr, and traps it in an inactive monomeric state (16). Here we describe the important considerations in library selection as well as the detailed steps in the high throughput fluorogenic screening assay.

2. Materials

2.1. Consumables and Equipment

1. Dimethyl sulfoxide (DMSO), 2-mercaptoethanol, and ethylenediaminetetraacetic acid (EDTA).
2. Round-bottom polypropylene 96-well assay plate and plate sealing tape.
3. A 25 ml reagent reservoir, multichannel pipets (12-channel), and disposable tips for 10 and 200 μl volumes.
4. Fluorescence microplate reader, table-top centrifuge with microplate rotors, and a small incubator.

2.2. Buffer Stocks

1. Prepare all aqueous solutions in MilliQ water followed by gentle shaking or mixing to ensure all solids have dissolved and all components have thoroughly mixed. Filter all 1 L stocks into sterile 1 L bottles. Store all reagent stocks at room temperature.

2. Prepare a 10 ml stock of 0.5 M EDTA by dissolving 2.9 g EDTA into 10 ml of water.
3. Prepare a 1 L stock of 1 M K_2HPO_4 by dissolving 141.9 g K_2HPO_4 into 1 L of water.
4. Prepare a 1 L stock of 1 M KH_2PO_4 by dissolving 136.09 g KH_2PO_4 into 1 L of water.
5. Prepare 1 L stock of 3 M KCl by dissolving 223.6 g KCl into 1 L of water.

2.3. Assay Buffer

1. The assay buffer consists of 25 mM potassium phosphate pH 8, 150 mM KCl, 0.1 mM EDTA, and 1 mM 2-mercaptoethanol.
2. Prepare a 100 ml stock of assay buffer by mixing 2.4 ml of 1 M K_2HPO_4, 0.15 ml of 1 M KH_2PO_4, 5 ml of 3 M KCl, 20 μl of 0.5 M EDTA, and 92.48 ml water. Adjust pH to 8.0 using HCl and NaOH. Before use add 7 μl of 2-mercaptoethanol.

2.4. Protease Expression and Purification

Recombinant KSHV Pr is expressed in *Escherichia coli* and purified as reported previously (16). 80 μM aliquots of purified protease are flash frozen in storage buffer (same as assay buffer described in Subheading 2.3) and stored at −20°C. The total amount of protein required will vary depending on the assay and the number of plates being screened.

2.5. Protease Substrate Synthesis and Purification

The protease substrate is an optimized hexa-peptide with the fluorogenic reporter group 7-amino-4-carbamoyl-methylcoumarin (ACC), which allows for monitoring enzyme activity spectroscopically. The peptide sequence is Ac-Pro-Val-Tyr-tBug-Gln-Ala-ACC with an observed K_M of 8.5 ± 0.8 μM for KSHV Pr. Substrate is synthesized using standard FMOC chemistry and purified as reported previously (7, 12). Substrate stocks of 10 mM in 100% DMSO are stored at −20°C. The total amount of substrate required will vary depending on the assay and the number of plates being screened. (Ac = acetyl group, tBug = *t*-butyl glycine).

2.6. Test Compounds

For general considerations on library selection (see Note 1). Compound libraries are generally provided in ready-to-use format, dissolved in 100% DMSO and plated in either 96-well or 384-well plates. An additional dilution step with 100% DMSO or plate reformatting may be necessary depending on the screening assay conditions. In order to avoid screening artifacts resulting from compound aggregation or precipitation, a final screening concentration between 10 and 30 μM is recommended. Our compound library was in a ready-to-use format at a concentration of 1 mM in a 96-well plate. Compound libraries are stored at −20°C.

2.7. Control Compound

Positive and negative controls are an important measure of assay performance and should ideally be included in every row (or column) of the assay plate. In assays, where timing is of particular importance, such as a protease assay, these controls allow for accurate calculation of % inhibition in each row (or column), as well as comparison of data across all rows (or columns). DMSO serves as the negative control and is used in the place of test compound. An ideal positive control may be a known active compound, such as a protease inhibitor in our case. The total amount of control compound required will vary depending on the assay and the number of plates being screened. At the time we performed our screen potent reversible KSHV Pr inhibitor did not exist, therefore substrate alone, which mimics the absence of protease activity, served as the positive control.

3. Methods

Prior to performing a high-throughput screen, it is critical to evaluate the quality of the screening assay by calculating a Z'-factor (see Note 2). High-throughput screening is typically performed in small assay volumes, therefore small fluctuations in liquid handling may affect the final readout. For this reason working with professionally calibrated pipettes is highly recommended. The following is a step-by-step protocol for manually performing a screen in one 96-well plate. The ability to screen multiple plates at once depends on the assay conditions and access to liquid handlers.

3.1. Summary of Assay Conditions

1. The protease assay is carried out in a final volume of 100 μl per well. It is composed of 98 μl of KSHV Pr in assay buffer, 1 μl of test compound or DMSO, and 1 μl of the protease substrate. These proportions ensure that the final concentration of protease is minimally changed upon substrate addition.
2. In order to account for volume loss due to pipeting, add an extra 20% to all calculated reagent volumes. For example, to calculate the total volume of protease substrate required for 40 wells use the equation: $[40 \text{ wells} + (0.2 \times 40 \text{ wells})] \times 1\ \mu l = 9.5$ ml.
3. KSHV Pr is used at a final concentration of 2 μM. We selected a concentration close to the reported in vitro K_D of 1.7 μM, in order to identify compounds that effectively compete with protease dimerization.
4. Test compounds are used at a final concentration of 10 μM. This concentration is typical for high throughput screening and helps avoid precipitation and aggregation.
5. Protease substrate is used at a final concentration of 100 μM, which is roughly ten fold above its reported K_M of 8.5 ± 0.8 μM

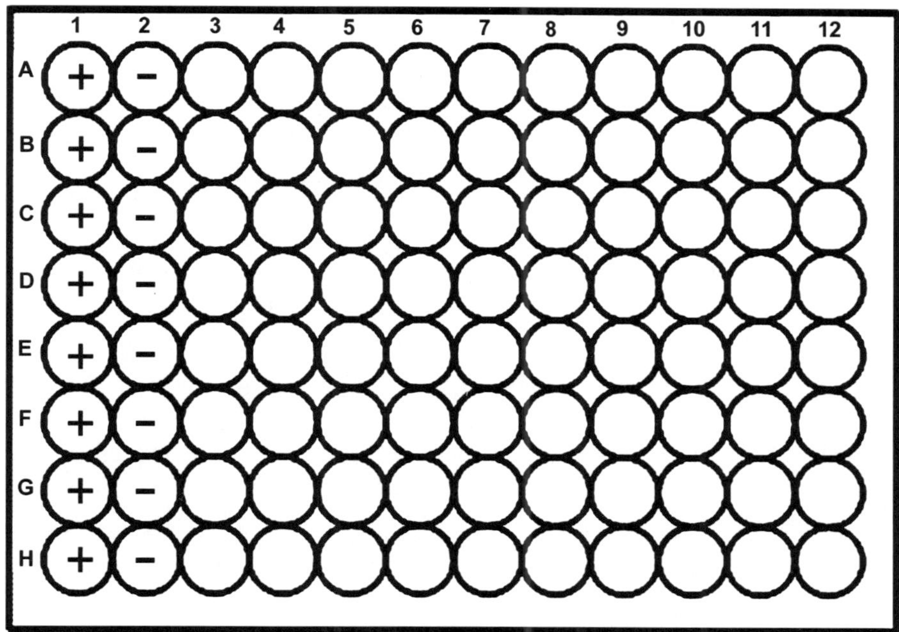

Fig. 4. The 96-well plate arrangement. Positive (+) and negative (–) controls are located in columns 1 and 2, respectively. Test compounds are screened in columns 3–12. The wells in each row receive substrate at the same time. The controls in each row may be used to accurately calculate % inhibition.

for KSHV Pr. This concentration ensures that substrate binding is not the rate-limiting step in the screening assay.

6. Both the protease substrate and test compounds are dissolved in 100% DMSO. Therefore the final assay concentration of DMSO is 1% (v/v), which is well below the previously determined 5% DMSO tolerance of the assay (see Note 3).

3.2. Setup of 96-Well Plate

1. Column 1 is designated for the positive control. The components are 98 μl of assay buffer, 1 μl of DMSO, and 1 μl of protease substrate (Fig. 4).
2. Column 2 is designated for the negative control. The components are 98 μl of protease in assay buffer, 1 μl of DMSO, and 1 μl of protease substrate.
3. Columns 3–12 are designated for the test compounds. The components are 98 μl of protease in assay buffer, 1 μl of test compound, and 1 μl of protease substrate.

3.3. Dispensing KSHV Pr to Assay Plate

1. Retrieve a frozen 80 μM vial of KSHV Pr from the freezer and thaw quickly by placing inside a 37°C water bath. Once the protease is thawed, store the vial on ice.
2. Transfer 10.5 ml of assay buffer into clean 50-ml conical tube.
3. Add 275 μl of KSHV Pr from the thawed stock and mix. Label this conical tube "KSHV Pr."

4. Carefully pour the entire mixture into a clean reagent reservoir labeled "KSHV Pr."
5. Retrieve a clean 96-well plate, label it "Assay Plate," and place on work bench at room temperature.
6. Load a 200 μl-volume multichannel pipette with eight clean tips, place inside "KSHV Pr" reservoir, and load 98 μl. Inspect all eight channels by eye to make sure the pipette is functioning properly. If large air bubbles are trapped inside the tip, empty all channels back into the "KSHV Pr" reservoir and try again.
7. Dispense contents into column 3, and then repeat until all remaining columns 4–12 receive KSHV Pr.

3.4. Dispensing Test Compounds to Assay Plate

1. Retrieve compound library from the −20°C freezer and thaw at room temperature. Do not remove the plate seal.
2. Spin down plate in a table-top centrifuge at 3,000 × *g* for 1 min. After centrifugation remove the seal, label the plate "Compound Plate," and place it on the work bench at room temperature.
3. Load a 10 μl-volume multichannel pipette with eight clean tips, place inside column 1 of "Compound Plate," and load 1 μl.
4. Dispense contents into column 3 of "Assay Plate," such that compounds from wells A1–H1 of "Compound Plate" are transferred to wells A3–H3 of the "Assay Plate."
5. Repeat steps 3–4 to transfer from columns 2–10 of "Compound Plate" to the corresponding columns 4–12 on the "Assay Plate."
6. Columns 11–12 of the "Compound Plate" may be screened in a separate 96-well assay plate.

3.5. Dispensing the Negative Control to Assay Plate

1. Load a 200 μl-volume multichannel pipette with eight clean tips, place inside "KSHV Pr" reservoir, and load 98 μl.
2. Dispense contents into column 2 of "Assay Plate."
3. In a fresh reagent reservoir, labeled "DMSO," add 20 ml of DMSO.
4. Load a 10 μl-volume multichannel pipette with eight clean tips, place inside the "DMSO" reservoir, and load 1 μl.
5. Dispense contents into column 2 of "Assay Plate."

3.6. Dispensing the Positive Control to Assay Plate

1. Add 20 ml of assay buffer into a clean reagent reservoir labeled "Assay Buffer."
2. Load a 200 μl-volume multichannel pipette with eight clean tips, place inside the "Assay Buffer" reagent reservoir, and load 98 μl.
3. Dispense contents into column 1 of "Assay Plate."

4. Load a 10 μl-volume multichannel pipette with eight clean tips, place inside "DMSO" reservoir, and load 1 μl.
5. Dispense contents into column 1 of "Assay Plate."

3.7. Incubation of Assay Plate

1. Load a 200 μl-volume multichannel pipet, with eight clean tips and set to 50 μl.
2. Place tips midway inside column 1 of "Assay Plate," load and empty three times to mix contents.
3. Repeat steps 1–2 for columns 2–12.
4. Cover the plate using a sealing tape and spin down in a table-top centrifuge at 3,000 × *g* for 1 min.
5. Place plate inside 30°C incubator for 30 min.

3.8. Addition of Protease Substrate to Assay Plate

1. Spin down "Assay Plate" in a table-top centrifuge at 3,000 × *g* for 1 min. Gently remove the plate seal.
2. Retrieve 10 mM stock of protease substrate from −20°C freezer and thaw at room temperature.
3. In a fresh 96-well plate, labeled "Substrate Plate," aliquot 15 μl of substrate into each well of row A.
4. Load a 10 μl-volume multichannel pipette with 12 clean tips, place inside row A of "Substrate Plate," and load 1 μl.
5. Dispense contents into row A of "Assay Plate."
6. Load a 200 μl-volume multichannel pipette with 12 clean tips and set to 50 μl.
7. Place tips midway inside row A of "Assay Plate," load and empty three times to mix contents.
8. Repeat steps 4–7 for rows B–H of "Assay Plate."

3.9. Incubation of Assay Plate

1. Cover the plate using a new sealing tape and spin down in a table-top centrifuge at 3,000 × *g* for 1 min.
2. Place plate inside 30°C incubator for 60 min. We have confirmed experimentally that the observed rate of substrate hydrolysis is linear during this time period.

3.10. Stopping the Enzyme Reaction

1. Spin down "Assay Plate" in a table-top centrifuge at 3,000 × *g* for 1 min.
2. Place "Assay Plate" on the work bench at room temperature. Carefully remove the sealing tape.
3. Load a 200 μl-volume multichannel pipette with 12 clean tips, place inside the "DMSO" reservoir, and load 100 μl.
4. Dispense contents into row A of "Assay Plate."
5. Load a 200 μl-volume multichannel pipette with 12 clean tips and set to 100 μl.

6. Place tips midway inside row A of "Assay Plate," load and empty three times to mix contents.
7. Repeat steps 2–5 for rows B–H of "Assay Plate."
8. Cover the plate with a new sealing tape.

3.11. Reading the Assay Plate in Fluorescence Plate Reader

1. Spin down "Assay Plate" in a table-top centrifuge at 3,000 × *g* for 1 min. Gently remove the plate seal.
2. Place in microplate reader and measure the endpoint fluorescence with the excitation and emission wavelengths set to 380 and 460 nm, respectively.

3.12. Data Processing

1. For each row corrected endpoint fluorescence using the equation: $\mathrm{EF} - C^{\mathrm{P}}$.
1. EF is the endpoint fluorescence of the experimental and negative control wells. C^{P} is the endpoint fluorescence of the positive control well.
2. For each row calculate the % inhibition using the equation: $1-(T^{\mathrm{EF}}/N^{\mathrm{EF}})$.
3. T^{EF} is the corrected endpoint fluorescence of the experimental well with test compound. N^{EF} is the corrected endpoint fluorescence of the negative control well.
4. Compounds that showed inhibition over 50% were considered true hits in our screen.
5. If possible repeat the screen and determine the average % inhibition from triplicate data set.
6. Follow-up experiments will rule out false positives and determine if hits act as dimer disruptors (see Notes 4–7).

4. Notes

1. KSHV Pr activity is regulated by a dimerization-driven conformational switch. The two α-helices located at the dimer interface became our rationale for screening a helical mimetic small-molecule library for inhibitors. α-Helices make up the largest class of protein secondary structure and are commonly found at the interface of protein–protein interactions (18, 19). Therefore significant effort has been aimed at developing non-peptide small molecule mimetics of α-helices. The key characteristic of such molecules is a rigid scaffold with functionalities presented in the same orientation as the *i*, *i*+3 or *i*+4, and *i*+7 residue positions on an α-helix. The chemotypes of commonly reported scaffolds include biphenyls, allenes, alkylidene cycloalkanes, spiranes,

benzylideneacetophenones, trisubstituted imidazole, indanes, polycyclic ethers, benzodiazepines, and teraryl units (18). We had access to a small focused library of helical mimetic small molecules synthesized to disrupt the interaction of p53 and MDM2, which also includes an α-helix (17). The scaffolds were made of two or three aryl rings connected by amide bonds, with side chains in the i, $i+4$, and $i+7$ positions. It is worth noting that commercial helical mimetic libraries, as well as general protein–protein interaction libraries, are now available for purchase through companies like BioFocus. In the absence of suitable small molecule test candidates, peptidic alternatives such as hydrocarbon-stapled helical peptides can also serve as a starting point (20–22).

2. The Z'-factor is the measure of assay quality and is vital to identifying true hits in a high throughput screen (23). The value is typically calculated from the positive and negative control experiments, which in our screen correspond to the "low" and "high" signals, respectively. The resulting numerical values range between 0 and 1, where numbers closest to 1 are favorable. The method of calculating the Z'-factor has been reported previously. Z'-factor optimization is achieved by varying the assay conditions such that the separation between the "low" and "high" signal is increased. We optimized our KSHV Pr assay by increasing the assay incubation temperature, the reaction incubation time, and including centrifugations steps.
3. DMSO exhibits an inhibitory affect in most assays, therefore its final concentration should be kept to a minimum. However, library compounds and protease substrates are generally dissolved in 100% DMSO stocks and often require some DMSO to remain dissolved in aqueous buffer. For this reason it is important to determine the DMSO tolerance of screening assays. This is done by performing the basic assay in the presence of increasing concentration of DMSO until an inhibitory condition is reached. The DMSO tolerance for the KSHV Pr assay is 5% (v/v).
4. Some compounds form aggregates that interact with proteins in a nonspecific manner (24, 25). Aggregate forming compounds may result in a false-positive inhibitory effect when screened as part of large libraries. Including a small amount of detergent, such as 0.01% Triton X-100 (v/v), will eliminate aggregation-based hits in most cases (25). In the case of assays like the KSHV Pr assay that don't tolerate detergents, including 1 mg/ml BSA eliminates nonspecific interactions. Ideally detergents are included in the initial screen. However, since the library we screened was small, we counter screened our hits in the presence of BSA.

5. Some compounds may exhibit fluorescent properties in the wavelengths monitored by the assay. For example a compound that absorbs light in the same region as our ACC substrate, may interfere with the final assay readout. Therefore, it is important to monitor the abortions and emission properties of all screening hits at the wavelengths used to make read measurements.
6. A true inhibitor has a sigmoidal dose–response curve with a hill slope value that is around 1. The concentration of inhibitor that achieves 50% inhibition is the IC_{50}. Dose–response curves are obtained by repeating the screening assay with increasing concentrations of compound, ranging from 0 to 100 μM. Generally a threefold dilution of compound is used (100, 33.3, 11.1, etc.). Percent inhibition values (see Subheading 3.11) are then plotted as a function of inhibitor concentration.
7. Since the goal of the screen is to identify dimer disruptors, secondary assays that monitor dimerization must be available. We used two independent but complementary methods to determine weather our hit DD2 was a dimer disruptor (7, 16). The first was an FPLC assay and the second was a 2D-NMR approach using protease containing a reporter at the dimer interface. Both methods were carried out as described previously.

Acknowledgments

This work was supported by NIH grants T32 GMO7810, AIO67423 (C.S.C.), and by the American Lebanese and Syrian Associated Charities and St Jude Children's Research Hospital (R.K.G.).

References

1. Borthwick AD et al (2002) Design and synthesis of pyrrolidine-5,5-trans-lactams (5-oxohexahydropyrrolo[3,2-b]pyrroles) as novel mechanism-based inhibitors of human cytomegalovirus protease. 2. Potency and chirality. J Med Chem 45:1–18
2. Borthwick AD et al (2002) Pyrrolidine-5,5-trans-lactams as novel mechanism-based inhibitors of human cytomegalovirus protease. Part 3: potency and plasma stability. Bioorg Med Chem Lett 12:1719–22
3. Borthwick AD et al (1998) Design and synthesis of monocyclic beta-lactams as mechanism-based inhibitors of human cytomegalovirus protease. Bioorg Med Chem Lett 8:365–70
4. Gopalsamy A et al (2004) Design and syntheses of 1,6-naphthalene derivatives as selective HCMV protease inhibitors. J Med Chem 47:1893–9
5. Ogilvie W et al (1997) Peptidomimetic inhibitors of the human cytomegalovirus protease. J Med Chem 40:4113–35
6. Waxman L, Darke PL (2000) The herpesvirus proteases as targets for antiviral chemotherapy. Antivir Chem Chemother 11:1–22
7. Marnett AB, Nomura AM, Shimba N, Ortiz de Montellano PR, Craik CS (2004)

Communication between the active sites and dimer interface of a herpesvirus protease revealed by a transition-state inhibitor. Proc Natl Acad Sci USA 101:6870–5
8. Nomura AM, Marnett AB, Shimba N, Dotsch V, Craik CS (2005) Induced structure of a helical switch as a mechanism to regulate enzymatic activity. Nat Struct Mol Biol 12:1019–20
9. Nomura AM, Marnett AB, Shimba N, Dotsch V, Craik CS (2006) One functional switch mediates reversible and irreversible inactivation of a herpesvirus protease. Biochemistry 45:3572–9
10. Pray TR, Nomura AM, Pennington MW, Craik CS (1999) Auto-inactivation by cleavage within the dimer interface of Kaposi's sarcoma-associated herpesvirus protease. J Mol Biol 289:197–203
11. Pray TR, Reiling KK, Demirjian BG, Craik CS (2002) Conformational change coupling the dimerization and activation of KSHV protease. Biochemistry 41:1474–82
12. Lazic A, Goetz DH, Nomura AM, Marnett AB, Craik CS (2007) Substrate modulation of enzyme activity in the herpesvirus protease family. J Mol Biol 373:913–23
13. Reiling KK, Pray TR, Craik CS, Stroud RM (2000) Functional consequences of the Kaposi's sarcoma-associated herpesvirus protease structure: regulation of activity and dimerization by conserved structural elements. Biochemistry 39:12796–803
14. Shimba N, Nomura AM, Marnett AB, Craik CS (2004) Herpesvirus protease inhibition by dimer disruption. J Virol 78:6657–65
15. Lee GM, Craik CS (2009) Trapping moving targets with small molecules. Science 324:213–5
16. Shahian T et al (2009) Inhibition of a Viral Enzyme by a Small Molecule Dimer Disruptor. Nat Chem Biol 9:640–6
17. Lu F et al (2006) Proteomimetic libraries: design, synthesis, and evaluation of p53-MDM2 interaction inhibitors. J Comb Chem 8:315–25
18. Cummings CG, Hamilton AD (2010) Disrupting protein-protein interactions with non-peptidic, small molecule alpha-helix mimetics. Curr Opin Chem Biol 14:341–6
19. Davis JM, Tsou LK, Hamilton AD (2007) Synthetic non-peptide mimetics of alpha-helices. Chem Soc Rev 36:326–34
20. Gavathiotis E et al (2008) BAX activation is initiated at a novel interaction site. Nature 455:1076–81
21. Henchey LK, Jochim AL, Arora PS (2008) Contemporary strategies for the stabilization of peptides in the alpha-helical conformation. Curr Opin Chem Biol 12:692–7
22. Moellering RE et al (2009) Direct inhibition of the NOTCH transcription factor complex. Nature 462:182–8
23. Zhang JH, Chung TD, Oldenburg KR (1999) A Simple Statistical Parameter for Use in Evaluation and Validation of High Throughput Screening Assays. J Biomol Screen 4:67–73
24. Feng BY, Shelat A, Doman TN, Guy RK, Shoichet BK (2005) High-throughput assays for promiscuous inhibitors. Nat Chem Biol 1:146–8
25. Feng BY, Shoichet BK (2006) A detergent-based assay for the detection of promiscuous inhibitors. Nat Protoc 1:550–3

Chapter 10

Use of a Fluorescent ATP Analog to Probe the Allosteric Conformational Change in the Active Site of the Protein Kinase PDK1

Valerie Hindie, Laura A. Lopez-Garcia, and Ricardo M. Biondi

Abstract

There is growing interest in exploring allosteric sites on proteins for drug discovery. At the center of the regulation of many protein kinases from the AGC family there is an allosteric site termed "PIF-pocket." The regulated binding of a C-terminal region of the kinase to the PIF-pocket, within the small lobe of the catalytic core, modulates the activity of AGC kinases. Small compounds that bind to the PIF-pocket can mimic its physiological mechanism of regulation and modulate the kinase activity in vitro, e.g., small compounds can activate the phosphoinositide-dependent protein kinase 1 (PDK1). Compounds binding to an allosteric site on a protein kinase may produce conformational changes at the ATP-binding site within the active site of the kinase domain. We here describe a fluorescent method using the ATP analog TNP-ATP that allows evaluating the allosteric conformational changes at the ATP-binding site of PDK1 triggered by small compounds binding to the PIF-pocket.

Key words: PDK1, AGC protein kinase, Conformational change, Allosteric site, ATP-binding site

1. Introduction

Approximately one-third of new drug development programs in pharmaceutical industry are targeting protein kinases for the development of inhibitors (1). Most protein kinase inhibitors to date target the ATP-binding site at the active center of protein kinases. This site is relatively conserved within the over 500 protein kinases encoded by the human genome. Therefore, ATP-competitive compounds are often not selective for one protein kinase leading to potential side effects of the therapy. Protein kinase regulatory sites and docking sites with substrates can be used as possible alternative

Yi Zheng (ed.), *Rational Drug Design: Methods and Protocols*, Methods in Molecular Biology, vol. 928,
DOI 10.1007/978-1-62703-008-3_10,

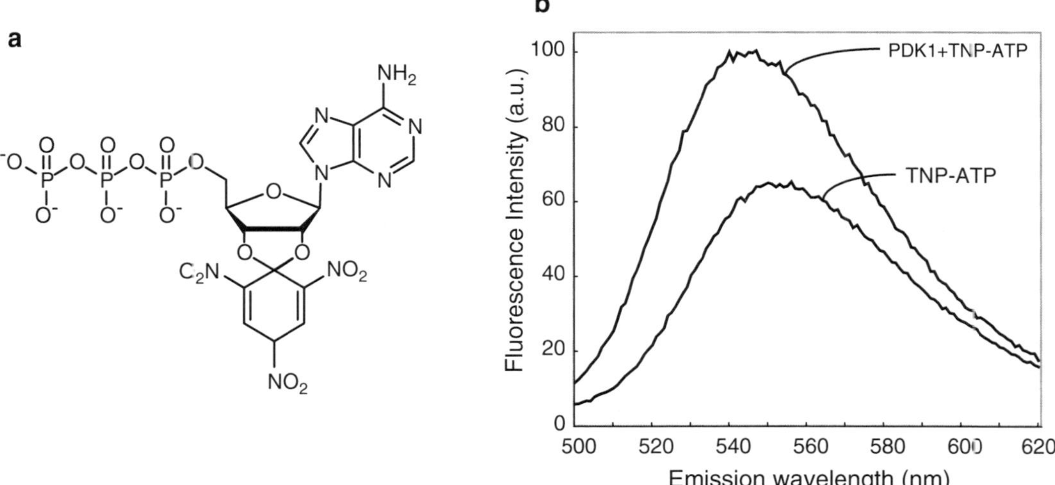

Fig. 1. The fluorescence of TNP-ATP increases upon interaction with PDK1. (**a**) Formula of TNP-ATP. (**b**) Fluorescence signal emitted by TNP-ATP (40 μM) alone or in the presence of PDK1 (15 μM). Reproduced with permission from ref. 5.

sites for non-ATP-competitive drugs (2–4) to improve the selectivity of a compound for a target enzyme. Allosteric compounds bind to sites different from the ATP-binding site and promote conformational changes of the enzyme. Interestingly, targeting allosteric sites can lead not only to allosteric inhibitors but also to allosteric activators of enzymes. Thus, there is growing interest in methods to characterize the effect of small compounds on the conformation of protein kinases.

AGC kinases possess a regulatory allosteric site (termed PIF-pocket) in the small lobe of the protein kinase core that can be targeted by small compounds. Binding of low-molecular-weight compounds to the PIF-pocket of the phosphoinositide-dependent protein kinase 1 (PDK1) activates the kinase. Thus, it was of interest to investigate the conformational change that prompted this activation. We set up an assay using the fluorescent ATP analog, 2′,3′-*O*-Trinitrophenyl-ATP (TNP-ATP) (Fig. 1a) to monitor changes at the ATP-binding site of protein kinases upon incubation with compounds that interact with the PIF-pocket allosteric site (5). In practical terms, we probed the conformation of the ATP-binding site by scanning the steady-state fluorescence of TNP-ATP in different conditions, i.e., using a range of peptides and small-molecular-weight compounds binding to the defined allosteric site. This method can be used to detect changes at the ATP-binding site using compounds previously characterized by other techniques to target an allosteric site. In our case, we confirmed the binding site of the compounds by crystallography and hydrogen–deuterium exchange experiments (5) (see Note 1).

The fluorescence intensity of TNP-ATP increases with the addition of PDK1 and is competed with the addition of excess ATP, indicating that the TNP-ATP fluorescence intensity increased upon binding to the ATP-binding site on PDK1. Notably, the TNP-ATP fluorescence intensity decreases with the addition of PS08, a compound that specifically binds to the PIF-pocket of PDK1 and activates the kinase. In the same assay, the isomer compound of PS08, PS133 that does not bind to PDK1, did not affect the fluorescence intensity of TNP-ATP, indicating that the effect produced by PS08 was highly specific. Altogether, the method allowed us to confirm that the binding of activating molecules targeting the PIF-binding pocket of PDK1 produced an allosteric effect that could be sensed at the ATP-binding site.

2. Materials

Prepare all solutions using ultrapure Milli-Q water. TNP-ATP should be protected from light. All materials should be kept on ice except the compounds dissolved in DMSO which should remain at room temperature.

2.1. Equipment

Spectrofluorometer (e.g., Varian Cary Eclipse); 50 μl quartz cuvette.

2.2. Stock Solutions

10 mM TNP-ATP (Jena Bioscience), high quality purified protein kinase of interest (e.g., $PDK1_{50\text{-}359}$[Y288A, Q292A] 45 μM) (see Note 2), 1 M Tris–HCl pH 7.5, 1 M DTT, 0.2 M EDTA, 100% DMSO.

2.3. Reagents

1. Buffer A: 50 mM Tris–HCl pH 7.5, 0.1 mM EDTA, 1 mM DTT. Mix 2.5 ml 1 M Tris–HCl pH 7.5, 50 μl 1 M DTT, 25 μL 0.2 M EDTA and complete with Milli-Q water to reach 50 ml. Prepare enough volume of buffer A to do all the experiments and store on ice. Usually 50 ml of buffer A is more than enough since each sample is prepared in a total volume of 60 μl for each condition.
2. Buffer B: 4% DMSO, 50 mM Tris–HCl pH 7.5, 0.1 mM EDTA, 1 mM DTT. Add 40 μl of DMSO to 960 μl of buffer A to obtain a final concentration of 4% DMSO. This is the buffer used to prepare dilutions of the compounds to be tested and for the control without compounds. It is recommended to prepare the buffer in excess in order to avoid variations in DMSO concentration during the experiment.
3. Buffer C: 0.5 M NaCl, 50 mM Tris–HCl, 0.1 mM EDTA, 1 mM DTT. This is the buffer of the purified protein. The

concentration of the enzyme is 3× the desired final concentration in the assay. For example, PDK1 was used at a final concentration of 15 μM. Thus, a stock concentration of 45 μM of PDK1 was prepared in excess in the same buffer of the purified protein (the exact buffer from the last purification step). The PDK1 stock is kept on ice for all experiments. For each assay, 20 μl of the enzyme stock is added in a total volume of 60 μl (see Note 3).

4. Compounds: Chemical compounds are usually dissolved in 100% DMSO. In this protocol, the stock of compound (100% DMSO) must be diluted 1:25 in buffer A to obtain a final concentration of 4% DMSO (the same DMSO concentration as in buffer B) (see Note 4). Further dilutions of compounds are prepared by dilution in buffer B. For the assay, 5 μl of compound at 4% DMSO is added for each condition in a total volume of 60 μl (final DMSO concentration in the assay: 0.3%).
5. TNP-ATP solution: Prepare a dilution 12× concentrated (480 μM) adding 4.8 μl of 10 mM TNP-ATP to 95.2 μl of buffer A and keep on ice (see Note 5). The final concentration of TNP-ATP in the assay is 40 μM. Therefore, add 5 μl of 480 μM TNP-ATP in a total volume of 60 μl for each condition. It is recommended to prepare the TNP-ATP solution in excess in order to avoid variations between different TNP-ATP preparations during the same experiment.

3. Methods

All reactions will be done in a final volume of 60 μl by adding:

20 μl of the enzyme 3× concentrated (or buffer C as control).

5 μl of 12× concentrated TNP-ATP (or buffer A as control).

5 μl of compound in 4% DMSO (or buffer B as control).

30 μl of buffer A.

Under the described conditions, the final composition of the assay is:

50 mM Tris–HCl pH 7.5, 0.1 mM EDTA, 1 mM DTT, 167 mM NaCl, 0.3% DMSO, 40 μM TNP-ATP (when appropriate), 15 μM PDK1 (when appropriate).

For each condition, mix all reagents following the same order in an Eppendorf tube and mix gently using a pipette, avoiding the formation of bubbles. Incubate 5 min at room temperature. Transfer the total volume to the quartz cuvette (see Note 6) and proceed with the measurement (see Note 7).

3.1. Data Measurement Parameters

Data were obtained in a Varian Cary Eclipse spectrofluorometer (excitation at $\lambda = 479$ nm; emission scanning from $\lambda = 500$–600 nm; excitation slit = 10 nm; emission slit = 10 nm) at a rate of 200 nm/min, with 150 data points/100 nm scanning and 0.3 s averaging time. The maximum intensity of emission was obtained at a wavelength between 545 and 549 nm. The assay is performed under controlled temperature (20°C). Several controls have to be performed to ensure the feasibility of the experiment with small compounds (Subheading 3.2).

3.2. Controls

1. Measurement of the emission wavelength of the enzyme. Verify that the protein kinase of interest does not emit fluorescence at the same maximum wavelength as the TNP-ATP alone (Subheading 3.3). To this end, prepare the following mixture and measure the fluorescence as described above (Subheading 3.1):

   20 μl of 3× concentrated enzyme.

   5 μl of buffer B.

   35 μl of buffer A.

2. Measurement of the fluorescence of the compounds to be tested in the presence of TNP-ATP without the target protein kinase. The compounds must not interfere with the signal of TNP-ATP alone. Prepare the following mixture and measure the fluorescence as described above (Subheading 3.1):

   20 μl of buffer C.

   5 μl of 12× concentrated TNP-ATP.

   5 μl of compound in 4% DMSO.

   30 μl of buffer A.

3.3. Measurement of the Fluorescence of TNP-ATP

Under the conditions chosen for the assay and in the absence of the protein kinase, determine the wavelength to obtain the maximum intensity of emission of TNP-ATP. This wavelength will be used to compare the increase or decrease of the fluorescence in the presence of the protein kinase and compounds. Prepare the following mixture and measure the fluorescence as described above (Subheading 3.1):

20 μl of buffer C.

5 μl of 12× concentrated TNP-ATP.

5 μl of buffer B.

30 μl of buffer A.

Under the conditions described above, TNP-ATP produced approximately 1.5 arbitrary units (a.u.) of fluorescence, the maximum intensity of emission was obtained at a wavelength between 545 and 549 nm (Fig. 1b).

3.4. Measurement of the Fluorescence of TNP-ATP in the Presence of the Protein Kinase of Interest

The variation of fluorescence intensity will depend on the chemical nature of the site targeted by the TNP-ATP. In our case, we expect that the environment in the active conformation of the active site will not differ much between protein kinases. However, it is possible that TNP-ATP may bind to the inactive form of the active site, which may differ between different kinases. Thus, it is important to verify that there is a significant change of the fluorescence of TNP-ATP in the presence of the protein kinase of interest (Fig. 1b). Prepare the following mixture and measure the fluorescence as described above (Subheading 3.1):

20 μl of the enzyme 3× concentrated.

5 μl of 12× concentrated TNP-ATP.

5 μl buffer B.

30 μl of buffer A.

In our essay, the inclusion of the enzyme PDK1 increased the fluorescence intensity between 545 and 549 nm from 1.5 a.u. to approximately 3.3 a.u (considered 100% fluorescence intensity; Fig. 1b) (see Note 8).

3.5. Measurement of the Background

Run only a mixture of the buffer A (35 μl), buffer B (5 μl), and buffer C (20 μl) to obtain the background due to the buffers alone. For the analysis of the data, subtract the background to the values obtained.

3.6. Measurement of the Fluorescence Emitted by TNP-ATP in the Presence of the Enzyme and Allosteric Compounds

Prepare the following mixture and measure the fluorescence as described above (Subheading 3.1):

20 μl of 3× concentrated enzyme.

5 μl of 12× concentrated TNP-ATP.

5 μl of compound in 4% DMSO.

30 μl of buffer A.

Our crystallography and deuterium exchange work showed that PS08 binds to the PIF-pocket allosteric site on PDK1 (5). In the established assay, PS08 prompted a concentration-dependent decrease of the fluorescence intensity of TNP-ATP (Fig. 2a). As a control, the inactive isomer compound PS133, that does not bind nor affect the activity of PDK1, did not prompt the decrease in TNP-ATP fluorescence (Fig. 2b). Therefore, this control confirms that the decrease in TNP-ATP fluorescence is specific due to the binding of the compound and activation of the kinase (see Note 9).

3.7. Probe the Specific Binding of TNP-ATP to the Kinase of Interest by Adding Excess of ATP

As a control for the targeted site, inclusion of an excess of ATP (1 mM) significantly diminished the emission of fluorescence, suggesting that ATP competed with TNP-ATP for binding to PDK1, presumably at the ATP-binding site. Thus, in the presence of an excess of ATP, TNP-ATP leaves the site to be replaced with ATP

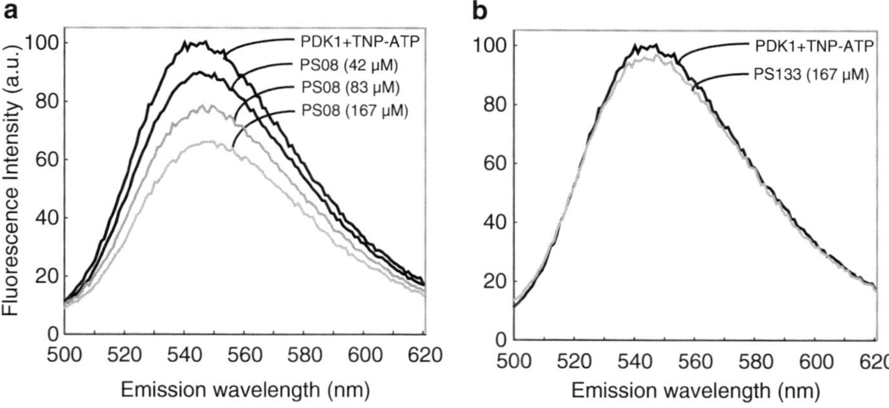

Fig. 2. Fluorescence of TNP-ATP in the presence of PDK1 is displaced by small-molecular-weight compounds that specifically interact with the PIF-pocket of PDK1. (**a**) Concentration-dependent displacement of TNP-ATP using the allosteric compound PS08. (**b**) Isomer PS133 that does not activate or interact with PDK1 does not displace the TNP-ATP signal. Reproduced with permission from ref. 5.

and therefore TNP-ATP emits similarly to the condition where it is alone in solution. Prepare the following mixture and measure the fluorescence as described above (Subheading 3.1):

20 μl of 3× concentrated enzyme.

5 μl of 12× concentrated TNP-ATP.

5 μl 4% DMSO buffer.

5 μl of 12 mM ATP (diluted in buffer A).

25 μl of buffer A.

As an additional control, it is recommended to verify that ATP has no effect on the fluorescence of TNP-ATP in the absence of a kinase. To this end, measure the fluorescence of a mix composed of TNP-ATP and ATP without an enzyme. Prepare the following mixture and measure the fluorescence as described above (Subheading 3.1):

20 μl of buffer C.

5 μl of 12× concentrated TNP-ATP.

5 μl 4% DMSO buffer.

5 μl of 12 mM ATP (diluted in buffer A).

25 μl of mother buffer.

3.8. Analysis of the Data

Each measurement should be done in duplicates. For the analysis of the data, make the average of each duplicated data point, subtract the background (obtained in Subheading 3.5), and draw the curve as arbitrary units (a.u.) versus emission wavelength or related to percentage of signal versus emission wavelength as in Figs. 1 and 2. It is recommended to repeat the measurement of the TNP-ATP alone and in the presence of the enzyme at the end of the set of experiments to confirm the stability of the reagents.

4. Notes

1. The method does not discriminate compounds that bind to an allosteric site from compounds that bind to the ATP-binding site displacing the TNP-ATP resulting in a decrease of the signal. The TNP-ATP binds to an inactive form of the protein kinase and the example here shows that compounds that allosterically activate the kinase displace TNP-ATP from the active site. It is expected that allosteric compounds that inhibit the kinase may promote the increase of TNP-ATP fluorescence.
2. The method described here was established using PDK1 catalytic domain (50–359) comprising two mutations (Y288A, Q292A) to compare our results to the crystal structure data that was obtained with this double mutant form of PDK1. The method was also used to monitor changes at the ATP-binding site of non mutated forms of PDK1 (50–359). The method can potentially be applied to other protein kinases or enzymes that bind ATP considering the necessary controls described in Subheading 3. The ideal concentration of enzyme and TNP-ATP would have to be experimentally determined for each new case.
3. The concentration of the enzyme needed for this method is rather high (above 1.5 mg/ml) and therefore can be rather viscous. To avoid pipetting errors we do not pipette small volumes of enzyme but rather dilute the stock of enzyme for the whole set of daily experiments at 45 μM and pipette 20 μl of enzyme into each assay.
4. Higher DMSO concentrations may increase the background. Depending on the protein tested, the authorized DMSO concentration may vary. The activity of protein kinases is usually not significantly affected by 1% DMSO.
5. It is recommended to prepare aliquots of the TNP-ATP stock to avoid repeated cycles of freezing and thawing. Keep the aliquots at −20°C in the dark until use.
6. We avoid the formation of bubbles in the cuvette by adding the sample at the bottom of the cuvette. If bubbles are formed when transferring the sample, gently tap the cuvette to help bubbles go to the surface.
7. Before starting the experiment, wash the cuvette with Milli-Q water several times and once with EtOH. After each measurement, rinse the cuvette twice with water and once with EtOH. It is important to dry the cuvette before adding each sample to keep the same concentrations for each measurement. To this end, use compressed air.

8. The physiological substrate of protein kinases is ATP-Mg. Inclusion of 2.5 mM $MgCl_2$ in the assay mix produced similar results. Therefore we excluded $MgCl_2$ from the assay, to avoid hydrolysis of TNP-ATP during the experiment. TNP-ADP can also be used instead of TNP-ATP, although the fluorescence intensity of TNP-ADP in the presence of PDK1 did not increase as much as using TNP-ATP.
9. The use of isomer compounds to probe the specificity of the assay may not be readily available for most new assays to be developed. In such case, related nonactive compounds may be used as a control.

References

1. Cohen P (2002) Protein kinases–the major drug targets of the twenty-first century? Nat Rev Drug Discov 1:309–315
2. Bogoyevitch MA, Fairlie DP (2007) A new paradigm for protein kinase inhibition: blocking phosphorylation without directly targeting ATP binding. Drug Discov Today 12:622–633
3. Kirkland LO, McInnes C (2009) Non-ATP competitive protein kinase inhibitors as antitumor therapeutics. Biochem Pharmacol 77:1561–1571
4. Lewis JA, Lebois EP, Lindsley CW (2008) Allosteric modulation of kinases and GPCRs: design principles and structural diversity. Curr Opin Chem Biol 12:269–280
5. Hindie V, Stroba A, Zhang H, Lopez-Garcia LA, Idrissova L, Zeuzem S, Hirschberg D, Schaeffer F, Jorgensen TJD, Engel M, Alzari PM, Biondi RM (2009) Structure and allosteric effects of low molecular weight activators on the protein kinase PDK1. Nat Chem Biol 5:758–764

Chapter 11

Affinity Purification of Protein Kinases that Adopt a Specific Inactive Conformation

Pratistha Ranjitkar and Dustin J. Maly

Abstract

Several protein kinases have been characterized in a specific inactive form called the DFG-out conformation. Unlike the active conformation which is conserved in all kinases, the inactive DFG-out conformation appears to be accessible to only certain kinases. This inactive conformation has been successfully targeted with highly selective kinase inhibitors, including the cancer drugs imatinib and sorafenib. However, the structural and sequence requirements for adopting this conformation are still poorly understood. Here, we describe a general method for enriching DFG-out adopting kinases from cell lysates with an affinity resin that contains a general ligand that specifically recognizes this inactive form.

Key words: Protein kinase, DFG-out conformation, Affinity purification, Inhibitors, Protein conformation, Protein ligands

1. Introduction

There are over 500 protein kinases encoded by the human genome (1). Through phosphorylation of substrates, kinases regulate important cellular processes such as growth, differentiation, and motility (2). Given their roles in numerous cellular functions, misregulation of kinases has been implicated in a number of diseases (3). Consequently, targeting kinases with small molecule therapeutics has gained considerable attention (3).

Two common types of small molecule inhibitors that target the ATP-binding sites of protein kinases have been described: Type I and Type II inhibitors (4). Type I inhibitors are ATP-competitive and can bind to the active form of kinases, whereas Type II inhibitors occupy an additional hydrophobic pocket that is only present in a specific inactive conformation (4). This conformation is called

Yi Zheng (ed.), *Rational Drug Design: Methods and Protocols*, Methods in Molecular Biology, vol. 928,
DOI 10.1007/978-1-62703-008-3_11, © Springer Science+Business Media New York 2012

"DFG-out" and is characterized by an almost 180° rotation of the conserved Asp-Phe-Gly (DFG) motif with respect to the active form (4, 5). Several highly selective inhibitors that target this state have been developed. However, the sequence and structural determinants that allow certain kinases to adopt this conformation are still largely unknown. Therefore, reagents that allow the identification of DFG-out adopting kinases are valuable tools for understanding the conformational dynamics of this important enzyme family.

We have recently reported a series of Type II inhibitors known as DSA compounds that appear to be general ligands for kinases that can adopt the inactive DFG-out conformation (6, 7). With this inhibitor scaffold, we have generated an affinity matrix that can be used to enrich DFG-out adopting kinases selectively from complex cellular mixtures (7).

Affinity purification with resin-bound small molecules has been used to determine the kinase selectivity of compounds in cellular lysates. By immobilizing the target compound on a solid support, and incubating it with cellular lysate, an unbiased kinase selectivity screen can be performed (8–11). A modified form of a general ligand for the DFG-out conformation (compound 1; Fig. 1) was used to generate Affinity Probe 1 (AP 1; Fig. 1). AP 1 was used to generate an affinity matrix that allowed an unbiased profiling of cellular lysates for kinases that are able to adopt the DFG-out conformation. (7). Additionally, a control probe, Affinity Probe 2 (AP 2; Fig. 1), which lacks the moiety that occupies the pocket created by movement of the DFG motif was also generated (7). This negative control allows the differentiation of proteins that bind nonspecifically to the resin and ensures only identification of kinases that are able to adopt this inactive conformation.

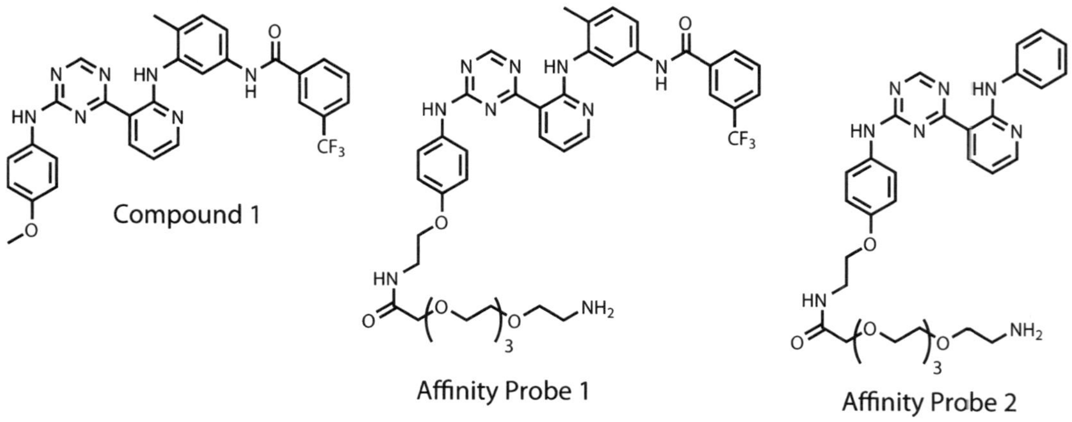

Fig. 1. Chemical structures of compound 1, Affinity Probe 1 AP 1, and Affinity Probe 2 AP 2.

2. Materials

Prepare all solutions using ultrapure water with a resistivity of 18 MΩ at 25 °C. All chemicals were purchased through commercially available sources unless otherwise noted. All chemicals should be handled with care and disposed of properly.

2.1. Affinity Matrix Generation

1. ECH Sepharose resin 4B.
2. 2 mL Handee Column.
3. AP 1, AP 2, and a nonlinkable internal standard, Compound 1, prepared according to Seeliger et al. (6) and Ranjitkar et al. (7).
4. Resin wash buffers: 0.5 M NaCl, pH 4.5 and 1:1 Dimethylformamide (DMF)–ethanol.
5. Affinity probe coupling reagents: Diisopropylethylamine (DIEA) (Sigma Aldrich; the reagent is used as provided by the supplier and the amount is specified in Subheading 3.1) and 1 M 1-ethyl-3-(3-dimethylaminopropyl) carbodiimide (EDCI) in 1:1 DMF–ethanol: Dissolve 383 mg of EDCI in 2 mL of 1:1 DMF–ethanol.
6. Quenching solution reagents: ethanolamine; 20 mM acetic acid, and 1 M EDCI prepared in 1:1 DMF–ethanol (as described in step 5).
7. High Performance Liquid Chromatography (HPLC) solvents: C_{18} (150 × 4.6 mm) analytical column; acetonitrile with 0.05% trifluoroacetic acid; water with 0.05% trifluoroacetic acid (see Note 1).

2.2. Lysate Preparation

1. Grow mammalian cells according to Ranjitkar et al. (see Note 2).
2. Lysis buffer: 50 mM HEPES (pH 7.5), 1 M NaCl, 0.5% Triton X-100, 1 mM EDTA, 1 mM EGTA, and 20 mM $MgCl_2$.
3. Protease inhibitor cocktail: Dissolve a tablet in 2 mL of H_2O to obtain a 25× stock (as indicated by the manufacturer).
4. 100 mM PMSF: Dissolve 17.4 mg in 1 mL of ethanol to make a 100 mM stock.
5. PD-10 Desalting Column.
6. Bradford Assay Reagent.

2.3. Affinity Pull-Down Experiments

1. Affinity Matrix 1 (AM 1; generated from AP 1) and Affinity Matrix 2 (AM 2; generated from AP 2). (Stored as a 30% slurry in 20% ethanol at 4 °C).
2. 10× Binding Buffer: 500 mM HEPES (pH 7.5), 5% Triton X-100, 10 mM EDTA, and 10 mM EGTA (see Note 3).
3. 6 M NaCl: Mix 350 mg NaCl in 1 mL H_2O.

4. Wash Buffer: 1× Binding Buffer, 150 mM NaCl. For example, to make 10 mL of wash buffer, mix 1 mL of 10× binding buffer, 250 μL of 6 M NaCl, and 8.75 mL water.
5. Elution Buffer: 1× Binding Buffer, 150 mM NaCl, 0.5% SDS. For example, to make 1 mL of elution buffer, mix 100 μL of 10× binding buffer, 25 μL of 6 M NaCl, 25 μL of 20% SDS in H_2O, and 850 μL water.
6. 3× SDS loading buffer: 240 mM Tris–HCl (pH 6.8), 6% (w/v) SDS, 30% glycerol, 2.3 M β-mercaptoethanol, and 0.06% (w/v) Bromophenol blue.
7. 10–20% Tris–HCl SDS-PAGE gel (Bio-Rad, Protean II Ready Gel, 160 × 160 × 1 mm).

2.4. Detection

1. Antibodies: Purchase commercially available antibodies as necessitated by the experiment.
2. SilverXpress Silver Staining Kit.

3. Methods

3.1. Generation of AM 1 and AM 2

Affinity matrix generation can be scaled according to the procedure described below.

1. Weigh out 2.25 mg of AP 1 (linkable probe) and 1.0 mg of Compound 1 (nonlinkable internal standard) in the same vial. Add 3.2 mL of 1:1 DMF–ethanol to prepare a 0.75 mM solution of probe AP 1 and 0.46 mM solution of Compound 1 (see Note 4).
2. Load 1.5 mL of ECH sepharose resin onto a 2 mL Handee column and let the storage buffer drain by gravity.
3. Wash the resin with 10 bed volumes (15 mL) of H_2O (pH 4.5; 0.5 M NaCl), then 1:1 DMF–ethanol.
4. Cap the column tip (see Note 5). Add 3 mL of the compound mix prepared in step 1. Then add 30 μL of DIEA and 450 μL of 1 M EDCI. Cap and parafilm the other end of the Handee column. Clamp the column onto an orbital rotator and incubate the coupling reaction for 24–48 h (see Note 6). Take the remaining 200 μL of the mix from step 1 and lyophilize the sample (see Note 7).
5. Resuspend the lyophilized sample in 20 μL acetonitrile. Analyze this sample with analytical HPLC (Fig. 2).
6. After 24–48 h, drain the coupling mix into a vial. Take 200 μL of the flow-through and lyophilize the sample. Resuspend the lyophilized sample in 20 μL acetonitrile and analyze the sample using analytical HPLC (Fig. 2).

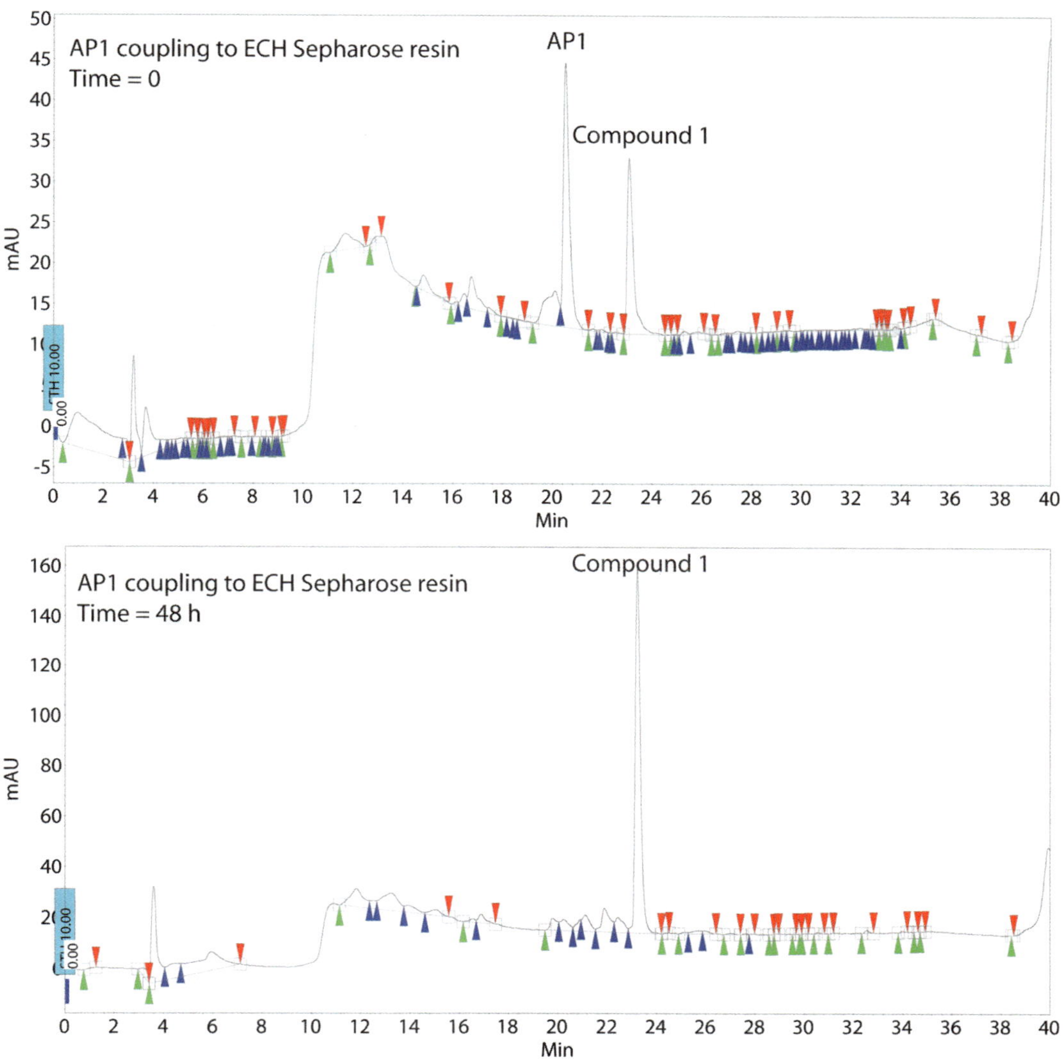

Fig. 2. AP 1 coupling to ECH sepharose resin as monitored by HPLC (UV absorption at 250 nm). The *top panel* shows the relative intensities of AP 1 and its nonlinkable counterpart, compound 1 (an internal standard), at time = 0. The *bottom panel* shows the flow-through collected from the resin after 48 h of incubation.

7. After the reaction is completed, wash the resin with 2 bed volumes (3 mL) of 1:1 DMF–ethanol by rotating the column for 5 min on an orbital rotator.
8. To quench the remaining reactive sites, incubate the resin with 56 μL of ethanolamine, 450 μL of 1 M EDCI and 960 μL of 20 mM acetic acid in 534 μL of 1:1 DMF–ethanol for 12 h.
9. Drain the reaction and discard the flow-through. Wash the resin with 6 bed volumes (9 mL) of 1:1 DMF–ethanol, 12 bed volumes (18 mL) of 0.5 M NaCl, and 10 bed volumes (15 mL) of 20% ethanol.

10. Add 3 mL of 20% ethanol to store the resin as 30% slurry at 4 °C until further use.
11. Prepare AM 2 following steps 1–10.

3.2. Lysate Preparation

1. Resuspend the cell pellet obtained from a 15-cm dish in 2 mL of lysis buffer, 20 μL of 100 mM PMSF and 80 μL of 25× Roche protease inhibitor cocktail in a 15-mL conical tube.
2. Using a 20 gauge needle and a 3 mL syringe, homogenize the lysate by gently pulling up the buffer into the syringe and pushing out at least five times. Avoid creating bubbles.
3. Place the Falcon tube on ice and sonicate (10 pulses at 50% duty cycle and 6.75 output control, 5×).
4. Clear the lysate by centrifuging at 16,000 × *g* for 60 min at 4 °C.
5. Desalt cleared lysate into lysis buffer using a PD-10 desalting column.
6. Determine the concentration of the lysate using a Bradford assay (see Note 8).
7. (Optional Step) Concentrate the lysate (4–6 mg/mL).
8. Snap-freeze the lysate and store at −80 °C until further use.

3.3. Affinity Pull-Down of Protein Kinases

Pull-down assays can be scaled according to the procedure below.

1. Thaw lysate prepared above on ice.
2. Transfer 150 μL of AM 1 and AM 2 into two separate 2 mL Handee columns. Wash both resins with 8 bed volumes of 1× wash buffer.
3. Remove 20 μL of the lysate and add 10 μL of 3× SDS loading buffer. This sample will be used as a control for total loaded protein.
4. Add 6–8 mg of lysate protein in a volume of 2 mL to AM 1 prepared in step 2.
5. Incubate the mixture for 2 h at 4 °C on an orbital rotator.
6. Drain the column and collect the flow-through. Remove 100 μL of the flow-through and add 50 μL of 3× SDS loading buffer for analysis.
7. Wash the resin with 8 bed volumes of wash buffer at 4 °C for 30 min on an orbital rotator. Repeat this step three times. For each wash, remove 100 μL of the sample and add 50 μL of 3× SDS loading buffer for analysis.
8. Add 4 bed volumes of elution buffer to the resin and incubate at room temperature for 10 min on an orbital rotator. Collect the elution sample. Repeat this step once.

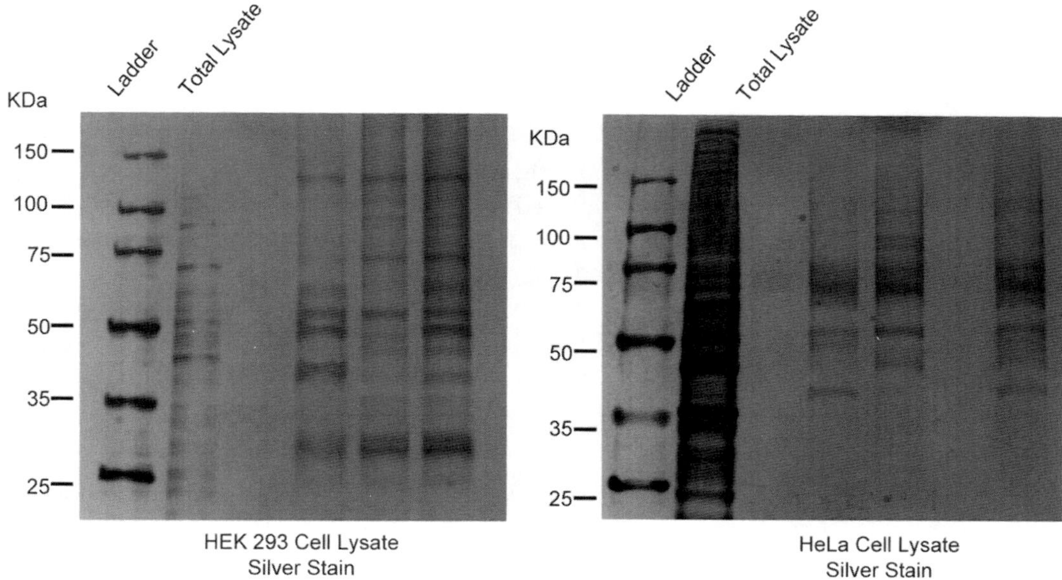

Fig. 3. *Left gel*: 8 mg of HEK-293 lysate was subjected to affinity purification with AM 1 (*lane 4*), AM 2 (*lane 5*), or AM 2 followed by AM 1 (*lane 6*). The first eluted fraction from each enrichment experiment was subjected to SDS-PAGE and silver stained. *Right gel*: 3 mg of HeLa lysate was subjected to affinity purification with AM 1 (*lane 4*), AM 2 (*lane 5*), or AM 2 followed by AM 1 (*lane 7*). The first eluted fraction from each enrichment experiment was subjected to SDS-PAGE and silver stained.

9. Lyophilize the elution samples and redissolve in 200 μL of 3× SDS loading buffer (see Note 9).
10. Repeat steps 4–9 with the AM 2 prepared in step 2. This affinity matrix services as a control for the identification of proteins that non-specifically bind to the resin.
11. Resolve all samples collected in steps 3, 6, 7, and 9 for AM 1 and AM 2 on a 10–20% Tris–HCl gradient gel (see Fig. 3).
12. Visualize the samples by silver staining or Western Blotting (see Note 10).

4. Notes

1. A C_{18} analytical column of any dimension can be used. The analytical method can be run in methanol or acetonitrile as long as the probe and the control compound have different retention times. We run samples in an acetonitrile–water mixture with a gradient of 1% to 100% acetonitrile over 40 min at a flow rate of 1 mL/min. Samples are monitored at 220 and 254 nm.

2. Ranjitkar et al. describes protocols for the preparation of HEK293 and HeLa cell lysates. Other cells lines can be used for these experiments as well.
3. 10× binding buffer can be stored long term at 4 °C.
4. The coupling efficiency of the affinity probe to the solid support is monitored by using analytical HPLC. Comparison of the relative ratios of the linkable probe versus its nonlinkable analog allows a quantitative assessment of the coupling efficiency. Therefore, it is best if the concentration of the nonlinkable internal standard is not allowed to exceed the concentration of the affinity probe as this prevents swamping out of the probe signal as it is coupled to the resin.
5. Parafilm the tip of the Handee column to prevent any leakage.
6. Monitor the coupling reaction by removing a small aliquot of the reaction mixture and analyzing it by analytical HPLC. If the ratio of AP 1 to compound 1 does not change with time, then the coupling reaction is complete. We have found that the reaction usually goes to completion within 24 h.
7. Lyophilizing the mixture is important because the sample can then be resuspended in a smaller volume and injected into the analytical HPLC. We have found that AP 1 and compound 1 are too dilute to be detected by HPLC without prior concentration of the samples.
8. Certain detergents interfere with the Bradford assay including Triton X-100, which is present in the lysis buffer. The Bradford reagent is compatible with up to 0.062% Triton X-100. Therefore, it is necessary to dilute the lysate ten-fold prior to Bradford analysis. Alternatively, one can use the BCA assay, which can tolerate up to 5% Triton-X 100.
9. In order to stay within the detection limit of silver staining, it is important to lyophilize the eluted samples and to resolubilize them in a smaller volume of SDS loading buffer. We have found that loading 60–65 μL of the enriched elution sample is enough to visualize bands with silver staining. One can also resolubilize the elution samples in a smaller volume to load the entire sample.
10. For visualization of specific proteins, run mini gels and perform a western blot. Probe for the target protein with specific antibodies.

Acknowledgment

This work was supported by the National Institute of General Medical Science (R01GM086858).

References

1. Manning G, Whyte DB, Martinez R, Hunter T, Sudarsanam S (2002) The protein kinase complement of the human genome. Science 298:1912–1934
2. Cohen P (2000) The regulation of protein function by multisite phosphorylation–a 25 year update. Trends Biochem Sci 25:596–601
3. Cohen P (2002) Protein kinases–the major drug targets of the twenty-first century? Nat Rev Drug Discov 1:309–315
4. Liu Y, Gray NS (2006) Rational design of inhibitors that bind to inactive kinase conformations. Nat Chem Biol 2:358–364
5. Okram B, Nagle A, Adrian FJ, Lee C, Ren P, Wang X, Sim T, Xie Y, Wang X, Xia G, Spraggon G, Warmuth M, Liu Y, Gray NS (2006) A general strategy for creating "inactive-conformation" abl inhibitors. Chem Biol 13:779–786
6. Seeliger MA, Ranjitkar P, Kasap C, Shan Y, Shaw DE, Shah NP, Kuriyan J, Maly DJ (2009) Equally potent inhibition of c-Src and Abl by compounds that recognize inactive kinase conformations. Cancer Res 69:2384–2392
7. Ranjitkar P, Brock AM, Maly DJ, Affinity reagents that target a specific inactive form of protein kinases. Chem. Biol. 17:195–206.
8. Knockaert M, Gray N, Damiens E, Chang YT, Grellier P, Grant K, Fergusson D, Mottram J, Soete M, Dubremetz JF, Le Roch K, Doerig C, Schultz P, Meijer L (2000) Intracellular targets of cyclin-dependent kinase inhibitors: identification by affinity chromatography using immobilised inhibitors. Chem Biol 7: 411–422
9. Knockaert M, Wieking K, Schmitt S, Leost M, Grant KM, Mottram JC, Kunick C, Meijer L (2002) Intracellular Targets of Paullones. Identification following affinity purification on immobilized inhibitor. J Biol Chem 277:25493–25501
10. Wissing J, Jansch L, Nimtz M, Dieterich G, Hornberger R, Keri G, Wehland J, Daub H (2007) Proteomics analysis of protein kinases by target class-selective prefractionation and tandem mass spectrometry. Mol Cell Proteomics 6:537–547
11. Godl K, Wissing J, Kurtenbach A, Habenberger P, Blencke S, Gutbrod H, Salassidis K, Stein-Gerlach M, Missio A, Cotten M, Daub H (2003) An efficient proteomics method to identify the cellular targets of protein kinase inhibitors. Proc Natl Acad Sci USA 100:15434–15439

Chapter 12

Determination of the Kinetics and Thermodynamics of Ligand Binding to a Specific Inactive Conformation in Protein Kinases

Sanjay B. Hari, Pratistha Ranjitkar, and Dustin J. Maly

Abstract

Recent interest in inactive kinase conformations has generated the need to develop new biochemical tools to study them. Here, we describe the use of a fluorescent probe that selectively and potently binds to a specific inactive conformation of protein kinases. This allows for the thermodynamics and kinetics of ligand binding to be determined.

Key words: Kinase, Inactive conformation, DFG-out, Activation loop, Fluorescent probe, Protein ligand, Dissociation constant

1. Introduction

Protein kinases represent approximately 1.7% of all human genes (1). This number is a testament to the vast array of kinase-mediated signal transduction pathways that regulate cellular processes such as immunity, division, and morphogenesis (2). Due in part to the success of kinase inhibitors such as Gleevec (imatinib) (3), there has been widespread interest in studying inactive kinase conformations. This information can be used to understand the regulation of this important family of enzymes and for targeted drug design. One commonly observed inactive kinase conformation, known as the "DFG-out" conformation, is characterized by a striking activation loop translocation and a revealed allosteric binding site (4).

Structure–activity relationship (SAR) studies using inhibitors that are known to recognize the inactive DFG-out form have been the primary method to determine which kinases can adopt this conformation. One caveat to these SAR studies is that they require kinases to be enzymatically active, rendering mutants or phosphoisoforms

Yi Zheng (ed.), *Rational Drug Design: Methods and Protocols*, Methods in Molecular Biology, vol. 928, DOI 10.1007/978-1-62703-008-3_12,

with limited activity incompatible with this approach. Despite the clear utility of fluorescent probes that target the ATP-binding sites of protein kinases, few have been reported. For example, the bicyclic imidazole inhibitor SK&F 86002 demonstrates an increase in fluorescence when complexed to the MAP kinase p38 (5), and a fluorescent analog of BIRB 796 has been used in p38 binding studies (6). However, these probes lack target generality and cannot be used to study other kinases. To this end, we have developed a small molecule fluorescent probe that relies on binding rather than activity to assess kinase active site conformation (7). Our probe is composed of a general ligand for the DFG-out conformation (8) conjugated to a BODIPY fluorophore. Upon binding to the active site of a protein kinase, this probe exhibits an increase in fluorescence.

Here we use this probe in a kinase titration assay to assess if a kinase is able to adopt the DFG-out inactive conformation. The experiment takes approximately 1 h from start to finish and yields direct binding data that would be otherwise difficult to obtain. Additionally, we describe a kinetic assay which can be used to determine how quickly the fluorescent probe conjugate dissociates from the kinase active site, providing insight into activation loop movement.

2. Materials

Prepare all buffers and solutions using ultrapure water with a resistivity of 18 MΩ at 25°C. The fluorescent probe is light sensitive and should always be handled in the dark (see Note 1).

2.1. Assay Components

1. Assay buffer (2×): 20 mM Tris–HCl (pH 7.5), 100 mM KCl, 1 mM $MgCl_2$, and 10% glycerol. Make 250 mL and sterilize using a 0.22 μM filter. Store at room temperature.
2. 10% Pluronic F-68 (Sigma).
3. 1 M Dithiothreitol (DTT): Dissolve 1.54 g in 10 mL water. Store at −20°C. Thaw frozen aliquots as needed.

2.2. Reagents

1. Kinase of interest: purified to >90% homogeneity as determined by SDS-PAGE (see Note 2).
2. Competitor C1 (Fig. 1, left structure): Prepare according to Seeliger et al. (8). Prepare a 2 mM solution in DMSO. Separate into aliquots and store at −20°C.
3. Fluorescent probe F1 (Fig. 1, right structure): Prepare BODIPY-conjugated C1 according to Ranjitkar et al. (7). Prepare a 2 mM solution in DMSO. Separate into aliquots and store at −20°C.

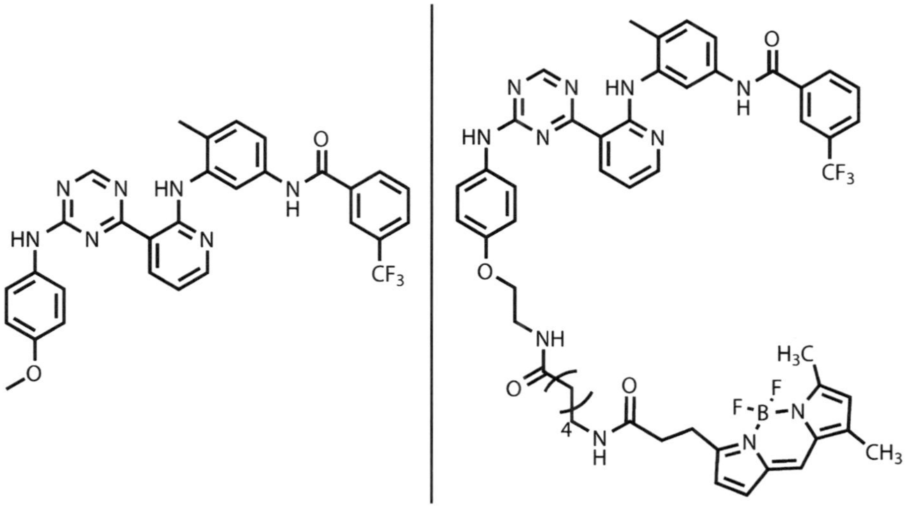

Fig. 1. Chemical Structures of competitor C1 (*left structure*) and fluorescence probe F1 (*right structure*).

2.3. Special Equipment and Consumables

1. Black, 96-well microplate (see Note 3).
2. Microplate reader capable of fluorescence excitation at 485 nm and emission detection at 535 nm (see Note 4).
3. 96-Well PCR microplates (see Note 5).

3. Methods

3.1. Affinity Measurements with Fluorescent Probe F1

1. Mix 2.25 mL of water, 2.25 mL of 2× assay buffer, and 4.5 μL of 1 M DTT in a 15-mL conical tube to yield a 1× assay buffer with 1 mM DTT.
2. Add 25 μL Pluronic F-68 to the 1× assay buffer prepared in step 1 and set aside.
3. Mix 1 μL of thawed fluorescent probe F1 (1 mM solution in DMSO) with 19 μL DMSO to make a 50 μM DMSO stock.
4. Mix 1.1 μL of the 50 μM fluorescent probe F1 solution prepared in step 3 with 53.9 μL of DMSO to yield a 1 μM working stock solution. The remaining 50 μM stock can be stored frozen at −20°C for later use (see Note 6). Cover the 1 μM fluorescent probe F1 solution and set aside.
5. Perform a twofold serial dilution of a protein kinase (preferred initial concentration = 10–15 μM) across one full row of a PCR microplate. Maintain the buffer composition of the kinase stock solution. Start with 80 μL of the initial protein kinase stock to yield 40 μL in each well.

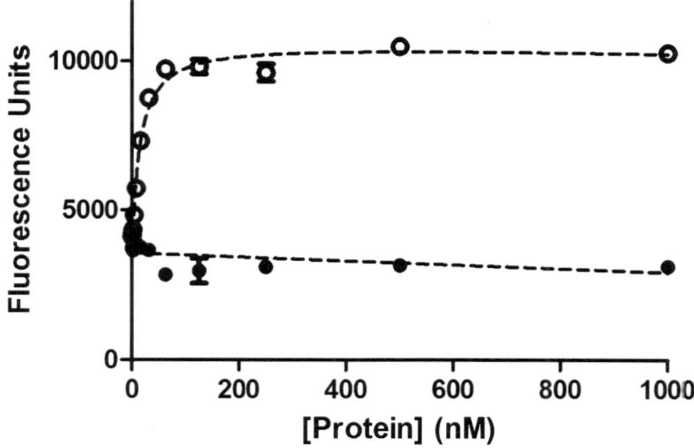

Fig. 2. Binding titration plots of p38α (*open circles*) and p38δ (*closed circles*) to fluorescent probe F1. These data represent a kinase that can (p38α) and cannot (p38δ) adopt the DFG-out inactive conformation. Data were fit using GraphPad Prism software to determine a dissociation constant (K_d) of 14 nM for p38α.

6. Transfer 10 μL of the kinase serial dilution prepared in step 5 to a black 96-well microplate with a multichannel pipettor.
7. Add 50 μL of the 1 μM fluorescent probe F1 solution prepared in step 4 to the 1× assay buffer prepared in step 2. Mix thoroughly. Dispense the mixture into a plastic reservoir.
8. Working quickly, add 110 μL of the probe mixture prepared in step 7 to each row of the black microplate with a multichannel pipettor (see Note 7). Incubate without agitation in the dark for 30 min at room temperature.
9. After 30 min, read the plate at Ex:485/Em:535 nm.
10. Use data analysis software (e.g., GraphPad Prism "Binding—Saturation, One site—Total") to determine the probe dissociation constant (Fig. 2) (see Note 8).

3.2. Kinetic Measurements of the Off Rate (k_{off}) and Dissociative Half-Life ($t_{1/2}$) of Fluorescent Probe F1

1. Mix 1 mL of water, 1 mL of 2× assay buffer, and 2 μL of 1 M DTT in a 15-mL conical tube to yield a 1× assay buffer with 1 mM DTT.
2. Add 0.6 μL of Pluronic F-68 to 109.8 μL of the 1× assay buffer prepared in step 1 and set aside.
3. Add 8.4 μL of a 10 μM kinase solution to the assay buffer prepared in step 2.
4. Mix 1 μL of thawed fluorescent probe F1 (1 mM solution in DMSO) with 19 μL of DMSO to make a 50 μM DMSO stock.
5. Mix 1.2 μL of the 50 μM fluorescent probe F1 solution prepared in step 4 to the reaction buffer prepared in step 3 [final kinase concentration = 700 nM; final fluorescent probe F1

concentration = 500 nM]. The remaining 50 μM DMSO stock of fluorescent probe F1 can be stored at −20°C for later use (see Note 6).

6. Incubate without agitation in the dark for 4 h at room temperature.
7. Shortly before the end of the incubation time, mix 886.5 μL of the 1× assay buffer solution prepared in step 1 with 4.5 μL Pluronic F-68 in a microcentrifuge tube.
8. Mix 12.5 μL of thawed competitor C1 (2 mM stock in DMSO) with 37.5 μL of DMSO to make a 500 μM competitor C1 DMSO stock.
9. Add 9 μL of the 500 μM competitor C1 stock prepared in step 8 to the assay buffer solution prepared in step 7 and mix thoroughly by pipetting.
10. At the end of the incubation period, dispense 10 μL of the reaction mixture prepared in step 5 to a well in a black microplate.
11. Add 290 μL of the competition solution prepared in step 9 to the microplate well containing 10 μL of the reaction mixture. Read immediately at Ex:485/Em:535, taking readings every 10 min for 8 h (see Note 9).
12. Use data analysis software (e.g., GraphPad Prism "Binding—Kinetics, Dissociation—One phase exponential decay") to determine the probe off rate (k_{off}) and dissociative half-life ($t_{1/2}$) (Fig. 3) (see Note 10).

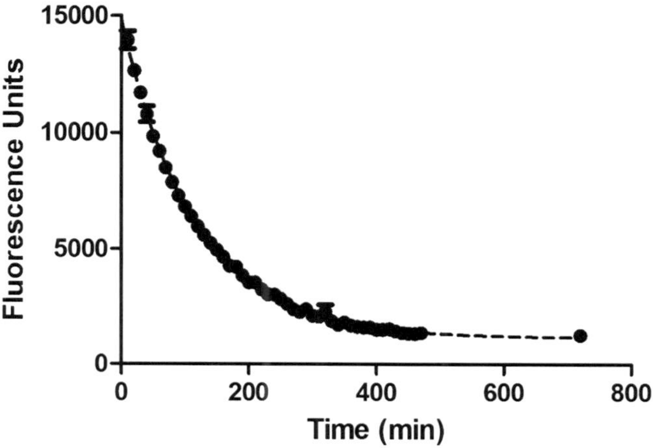

Fig. 3. Dissociation of fluorescent probe F1 from p38α. Data were fit using GraphPad Prism software to determine an off rate (k_{off}) of 161 μs^{-1} and a dissociative half-life ($t_{1/2}$) of 72 min.

4. Notes

1. Ambient light should be kept to a minimum whenever handling fluorescent probe stocks or solutions containing the fluorescent probe.
2. For best results, the kinase should be in a buffer free of imidazole and ionic detergents. Either desalting columns or dialysis membranes can be used for this purpose.
3. Presumably, any black microplate exhibiting low autofluorescence can be used. We use PerkinElmer OptiPlate-96F plates (#6005270) exclusively.
4. We use a PerkinElmer VICTOR3 Multilabel Plate Reader.
5. Any microplate can be used for serial dilutions, but we have found that low-volume, V-bottom plates such as those used for PCR are optimal for this task. We use those sold by Phenix Research (#MPS-499).
6. The quantum yield of the fluorescent probe diminishes significantly after multiple freeze-thaw cycles. Prepare separate aliquots of the fluorescent probe in DMSO to minimize the number of freeze-thaw cycles. In addition, include a positive control in your plate to confirm the integrity of the fluorescent probe.
7. We have found that aiming the pipette tips directly at the wells and rapidly dispensing obviates the need to mix the solutions.
8. Other data fitting programs such as QtiPlot can be used, but the equation used to fit the data must be entered manually. The binding equation used by GraphPad Prism is $Y=(B_{max} \times X)/(K_d+X)+NS \times X+$"Background," where X is protein concentration, Y is binding using (in this case fluorescence units), B_{max} is maximum binding, K_d is the dissociation constant, NS is the slope of nonspecific binding (assumed to be linear), and "Background" is binding in the absence of ligand. GraphPad Prism fits B_{max}, K_d, NS, and "Background" given a data set of X and Y. The "Background" variable can be constrained if its value is known: omit the 12th well during the titration in step 5, Subheading 3.1 to empirically determine background fluorescence.
9. If the probe dissociates quickly from the kinase (i.e., $t_{1/2}$ <2 min), it may be necessary to employ a negative control to use as the fluorescence reading at $t=0$ min. In this case, make a separate competition solution substituting DMSO for C1 and add to three additional wells.
10. Other data fitting programs such as QtiPlot can be used, but the equation used to fit the data must be entered manually. The kinetic equation used by GraphPad Prism is $Y=(Y_0-\text{NS})$

$\times \exp(-K \times X) + \text{NS}$, where X is time, Y is binding units (in this case fluorescence units), Y_0 is Y at time zero, NS is binding after a long time, and K is the rate constant. GraphPad Prism fits Y_0, NS, and K given a data set of X and Y. The variable Y_0 can be constrained if its value is known (see Note 9).

Acknowledgment

This work was supported by the National Institute of General Medical Science (R01GM086858).

References

1. Manning G, Whyte DB, Martinez R, Hunter T, Sudarsanam S (2002) The protein kinase complement of the human genome. Science 298:1912–1934
2. Manning G, Plowman GD, Hunter T, Sudarsanam S (2002) Evolution of protein kinase signaling from yeast to man. Trends Biochem Sci 27:514–520
3. Zimmermann J, Buchdunger E, Mett H, Meyer T, Lydon NB (1997) Potent and selective inhibitors of the Abl-kinase: phenylamino-pyrimidine (PAP) derivatives. Bioorg Med Chem Lett 7:187–192
4. Schindler T, Bornmann W, Pellicena P, Miller WT, Clarkson B, Kuriyan J (2000) Structural mechanism for STI-571 inhibition of abelson tyrosine kinase. Science 289: 1938–1942
5. Lee JC, Badger AM, Griswold DE, Dunnington D, Truneh A, Votta B, White JR, Young PR, Bender PE (1993) Bicyclic imidazoles as a novel class of cytokine biosynthesis inhibitors. Ann N Y Acad Sci 696:149–170
6. Pargellis C, Tong L, Churchill L, Cirillo PF, Gilmore T, Graham AG, Grob PM, Hickey ER, Moss N, Pav S, Regan J (2002) Inhibition of p38 MAP kinase by utilizing a novel allosteric binding site. Nat Struct Biol 9:268–272
7. Ranjitkar P, Brock AM, Maly DJ (2010) Affinity reagents that target a specific inactive form of protein kinases. Chem Biol 17:195–206
8. Seeliger MA, Ranjitkar P, Kasap C, Shan Y, Shaw DE, Shah NP, Kuriyan J, Maly DJ (2009) Equally potent inhibition of c-Src and Abl by compounds that recognize inactive kinase conformations. Cancer Res 69:2384–2392

Chapter 13

Purification and Specific Assays for Measuring APE-1 Endonuclease Activity

Adrian Esqueda, Mohammed Z. Mohammed, Srinivasan Madhusudan, and Nouri Neamati

Abstract

Human apurinic/apyrimidinic endonuclease-1 (APE-1) is essential for base excision repair and plays a major role in DNA repair and maintaining genomic stability. Cancer cells treated with conventional DNA-damaging agents develop resistance due in part to upregulation of enzymes involved in DNA repair. It is hypothesized that inhibiting DNA repair machinery should sensitize the cells to DNA-damaging agents. Previously, it has been shown that APE-1 is implicated in drug resistance and cancer progression. Therefore, APE-1 inhibitors are being sought after for their synergistic properties with various chemotherapeutics agents. Screening of several compound libraries and optimization of known inhibitors of APE-1 endonuclease activity have been accelerated by the use of high-throughput screening. Nevertheless, potential inhibitors must be tested in other counterscreens to validate their selectivity for APE-1. Here, we describe in-depth protocols for APE-1 purification and development of assays specific for APE-1 endonuclease activity.

Key words: Apurinic and apyrimidinic endonuclease, Base excision repair, Drug resistance, Purification, Assay development, Screening

1. Introduction

During the last decade the apurinic/apyrimidinic endonuclease-1 (APE-1) protein has gained attention as a potential target for chemotherapeutic agents for certain types of cancers (1). APE-1 is a multifunctional protein, participating in the BER of DNA repair, maintaining transcription factors in their active reduced state through its redox effector (Ref-1) function, and recently identified, endoribonuclease activity in c-myc mRNA (2). In the treatment of cancer, some chemotherapeutic strategies are aimed at deliberately damaging DNA and the replication process of DNA through the use of alkylating or oxidative agents and by the premature termination of DNA replication through the use of nucleoside analogs (3).

Yi Zheng (ed.), *Rational Drug Design: Methods and Protocols*, Methods in Molecular Biology, vol. 928, DOI 10.1007/978-1-62703-008-3_13, © Springer Science+Business Media New York 2012

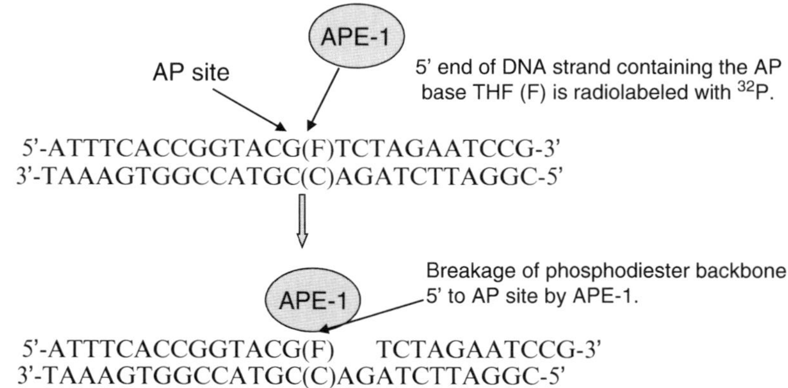

Fig. 1. Endonuclease activity of APE-1. During the base excision pathway (BER), APE-1 cleaves the phosphodiester backbone 5′ to the AP site. The 26-mer oligo used in this assay is radiolabeled with ^{32}P at the 5′ containing the AP site. The AP site in the 26-mer oligo is tetrahydrofuran (THF).

However, some cancers, such as gliomas and Acute Lymphocytic Leukemias (ACL), have demonstrated resistance to chemotherapeutic agents and radiation therapy (2, 3). Further investigation into the cause of the chemoresistance and radioresistance has revealed a strong link between elevated APE-1 protein expression and resistance to therapy, thereby legitimizing APE-1's candidacy as a chemotherapeutic target.

Identifying an inhibitor of the endonuclease activity of APE-1 has been challenging (Fig. 1). Highly guided searches of potential inhibitors through the screening of large compound libraries and pharmacophore-guided searches have focused the search for an inhibitor (1). High-throughput screening (HTS) assays of potential inhibitors for the endonuclease activity of APE-1 have been developed; however some of these assays require a large investment of capital in the purchase of highly sophisticated equipment based on technologies such as microfluidics (4–6). Despite the initial costs and findings of the HTS assays, the identified potential inhibitors must be tested in a gel-based assay for further validation. Using widely available and affordable materials and equipment, here we describe a PAGE-based assay in which a radiolabeled 26-mer DNA duplex with an abasic site is tested simultaneously with a potential inhibitor for the inhibition of APE-1's endonuclease activity.

2. Materials

All solutions should be prepared using sterilized and deionized water. All reagents used should be analytical grade reagents and unless specified otherwise, all reagents should be stored at room temperature.

2.1. Purification of Recombinant and His-Tagged APE-1 Enzyme Components

1. *E. coli* BL21(DE3) cells containing the pET28b-Ape1 vector.
2. LB broth MILLER media.
3. Isopropyl β-D-1-thiogalactopyranoside (IPTG).
4. 1× PBS: In 800 mL of sterilized and deionized water add 8 g of NaCl, 0.2 g of KCl, 1.44 g of Na_2HPO_4, and 0.24 g of KH_2PO_4. Bring volume to 1 L and adjust the pH to 7.4.
5. Phenylmethanesulfonylfluoride (PMSF), leupeptin, and pepstatin A.
6. Imidazole.
7. 2-Mercaptoethanol (BME).
8. Buffer A: 50 mM HEPES–KOH at pH 7.5, 0.5 M KCl, 10% glycerol, 0.1 mM EDTA, 0.1 mM DTT, and 1 mM imidazole.
9. Buffer B: 50 mM HEPES–KOH pH 7.5, 100 mM KCl, 10% glycerol, 1 mM EDTA, and 0.1 mM DTT.
10. Chromatography column with Swell-gel Nickel chelated disks (Pierce, Rockford, IL).

2.2. SDS-PAGE Gels, Buffers, and Staining Solutions for the Identification of Fractions with Recombinant APE-1 Protein

1. For the resolving section of the gel: 30% acrylamide, 1.5 M Tris pH 8.8, 10% SDS, 10% APS, and TEMED.
2. For the stacking section of the gel: 30% acrylamide, 0.5 M Tris pH 6.8, 10% SDS, 10% APS, and TEMED.
3. 10× Tris–Glycine SDS-PAGE running buffer for protein gel: Tris Base 30.3 g, Glycine 144 g, SDS 10 g, H_2O 1 L. Thoroughly mix together.
4. Gel casting apparatus and complementary gel running apparatus.
5. Destain solution 1: 50% EtOH, 7% acetic acid, 43% H_2O.
6. Destain solution 2: 5% EtOH, 7.5% acetic acid, 87.5% H_2O.
7. 5× SDS protein loading buffer dye: 10% SDS, 20 mM EDTA, 25% BME, 0.1% bromophenol blue, 50% glycerol.

2.3. Preparation of Radiolabeled Oligonucleotides

1. 1 mCi γ-P^{32}-labeled ATP.
2. Oligo substrate top strand with sequence: (5′ ATTTCACCG GTACG(F)TCTAGAATCCG 3′) with the tetrahydrofuran synthetic abasic residue. The top strand must be radiolabeled with γ-P^{32} using a T4 polynucleotide kinase. The complementary strand with sequence: (3′ TAAAGTGGCCATGC(C)AGA TCTTAGGC 5′).
3. T4 polynucleotide kinase (T4 PNK) and appropriate reaction buffer.

2.4. Optimization of APE-1 Recombinant Protein Concentration for Enzymatic Assay

1. APE-1 recombinant protein.
2. 100 mM $MgCl_2$ (5 mL).
3. BME (1:50 in water, 20 μL BME and 980 μL water).
4. Bovine Serum Albumin (BSA) 1 mg/mL.

5. 3-(*N*-morpholino) propanesulfonic acid (MOPS) at pH 7.2.
6. Glycerol.
7. Tris-base at pH 8.3.
8. EDTA.
9. 20% Acrylamide.
10. Urea.
11. Loading Dye and quencher for APE-1 enzymatic assay: 24.5 mL formamide, 73 mg EDTA, 6 mg xylene cyanol, 6 mg bromophenol blue.

2.5. Casting and Running of PAGE Gel

1. 10× TBE: 108 g of Tris base, 55 g of boric acid, 9.3 g EDTA, 800 mL H_2O. Bring volume to 1 L.
2. Denaturing 20% polyacrylamide gel: In a large bottle with a cap add and mix 30 g of urea, 30 mL 10× TBE, 30 mL of 2% *N*,*N*-methylene bis acrylamide solution, 75 g acrylamide, 75 mL H_2O (see Note 1). Mix thoroughly and filter the polyacrylamide mixture once it is in solution.
3. 10% Ammonium persulfate in water.
4. TEMED.
5. Apparatus for casting and running PAGE gel.
6. 3 mm Chromatography paper.
7. Saran wrap.
8. Exposure cassette and storage phosphor screen.
9. Storage phosphor Imager.

3. Methods

All procedures are to be carried out at room temperature unless specified otherwise and all material should be sterile before use.

3.1. γ-P^{32}-Labeled Oligo-Based APE-1 Endonuclease Cleavage Assay

3.1.1. Purification of Recombinant and His-Tagged APE-1 Enzyme

1. Initiate an overnight culture of *E. coli* BL21(DE3) cells containing the vector pET28b-Ape1.
2. Prepare 2 L of LB broth MILLER and supplement with 30 μg/mL of kanamycin.
3. Inoculate the 2 L of LB broth containing 30 μg/mL of kanamycin with a 10 mL aliquot of *E. coli* BL21(DE3) cells containing the vector pET28b-Ape1.
4. Incubate at 37°C with the shaking apparatus set at 250 revolutions per minute (rpm).
5. Induce protein expression by adding 400 μM IPTG during mid-log phase of bacterial growth with an approximate absorbance of 0.6–0.7 optical density at 595 nm (see Note 2).

6. Allow bacterial culture growth for 3 h at 37°C.
7. Collect the bacterial culture and centrifuge at 3,000 rpm for 20 min.
8. Wash the cell pellet once with 1× PBS and resuspend in 40 mL of buffer A supplemented with 1 mM PMSF, 2 μg/mL leupeptin, and 1 μg/mL of pepstatin A.
9. Lyse the cells by passing them twice through a French press (see Note 3).
10. Centrifuge the lysate at 30,000 × *g* for 30 min.
11. Prepare a Ni-affinity chromatography column with 2–3 Swell-gel Nickel-chelated discs.
12. Collect supernatant from step 10 and load onto the prepared Ni-affinity chromatography column from step 11.
13. Wash Ni-affinity chromatography column with 10 column volumes of buffer A supplemented with 1 mM imidazole.
14. Elute recombinant APE-1 protein with a linear increasing concentration of imidazole starting with 20 mM and ending at 100 mM (see Note 4).
15. Aliquot 20 μL of each postelution fraction and load on an SDS-PAGE gel to determine the fractions containing the recombinant APE-1 protein (see Note 5).

3.1.2. Casting and Running of SDS-PAGE Gels for Determination of Peak APE-1 Recombinant Protein Elution Fractions

1. Assemble the gel casting apparatus and prepare the gel running equipment.
2. In a 25 mL beaker mix together the components for the resolving gel and pour into the gel casting apparatus (see Note 6). Pour an additional 1 mL of water to flatten and level the resolving section of the gel. Allow 20 min for drying. Thereafter, remove the water carefully by tilting the gel casting apparatus.
3. In a 25 mL beaker mix together the components for the stacking gel, pour onto the remaining capacity of the gel casting apparatus, and insert the lane comb. Allow 20 min to dry.
4. For each postelution fraction of recombinant APE-1 protein, in a small tube mix 20 μL of the postelution recombinant APE-1 protein with 5 μL of 5× loading dye.
5. After the SDS-PAGE gel has dried and has been loaded onto the gel running apparatus and filled with 1× Tris–Glycine SDS-PAGE running buffer, load 20 μL of each recombinant APE-1 loading dye mix into the gel wells.
6. Run the gel at 100 V for 2 h.
7. After 2 h remove the gel and place it in an adequate container such that the gel can be completely submerged in destaining solution 1 (see Note 7). Once the gel has been submerged in destaining solution 1 microwave the gel for 1 min while still

submerged in the destaining solution 1. Pour off the destaining solution 1.

8. Using the same container, submerge the gel in destaining solution 2, add 2–3 droplets of Coomassie blue protein staining solution to the mixture, and microwave for 1 min. Pour off the destaining solution 2 and add a new volume of destaining solution 2 so as to submerge the gel again.
9. Within a few minutes (5–10 min) protein bands will become visible (see Note 8).
10. Once the postelution fractions have been identified as having strong APE-1 recombinant protein, pool the fractions together and store at −80°C.

3.1.3. Preparation of Radiolabeled Oligo Substrate

1. In a 1.5 mL microcentrifuge tube radiolabel top strand with sequence (5′ ATTTCACCGGTACG(F)TCTAGAATCCG 3′) by adding 10 μL of H_2O, 8 μL of 10× kinase buffer, 5 μL of the top strand oligo, 2 μL of T4 polynucleotide kinase (T4 PNK), and 45 μL of 1 mCi γ-P^{32}-labeled ATP. Incubate for 45 min at 37°C.
2. Heat the 1.5 mL microcentrifuge reaction tube to 85°C for 15 min (see Note 9).
3. To the same 1.5 mL microcentrifuge tube add 140 μL of H_2O and 10 μL of the complementary oligo strand (3′ TAAAGTGGCCATGC(C)AGATCTTAGGC 5′) and heat for an additional 2 min at 85°C. Thereafter, turn off the heat block and let the temperature slowly return to ambient temperature.
4. After prepping a Spin-25 mini-column, carefully pour the radiolabeled oligo duplex into the Spin-25 mini-column and centrifuge for 1 min at 2,000 rpm (see Note 10).

3.1.4. Optimization of Recombinant APE-1 Protein Concentration for Enzymatic Assay

1. Perform a BCA protein assay on the pooled peak postelution fractions of the recombinant APE-1 protein to determine the protein concentration.
2. In 4 separate 1.5 mL microcentrifuge tubes, label and make the following dilutions of the recombinant APE-1 protein (with a final volume 300 μL for each dilution) using water: 1:10, 1:100, and 1:1,000.
3. Prepare a template annotating the concentration of APE-1 protein in a decreasing trend (Fig. 2a). To identify the correct concentration of APE-1 recombinant protein, include a set of dilutions in which a known inhibitor of APE-1 is used in the experiment.
4. Label four 1.5 mL microcentrifuge tubes "1," "1:10," "1:100," and "1:1,000."
5. To each microcentrifuge tube in step 4 add 36 μL of 100 μM $MgCl_2$, 48 μL of 1 mg/mL BSA, 48 μL BME (diluted to 1:50

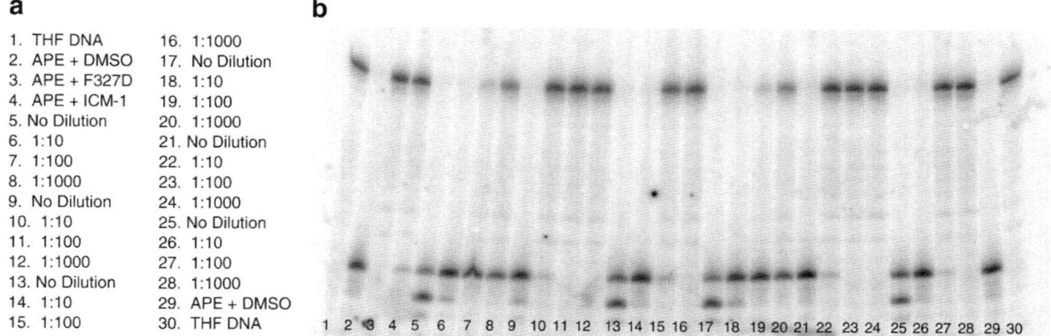

Fig. 2. Optimization of APE-1 recombinant protein assay and template. (**a**) Corresponding template for the optimization of APE-1 recombinant protein assay shown in (**b**). (**b**) The APE-1 recombinant protein used in this optimization assay is from the 600 mM imidazole elution fraction. Lanes 2, 3, 4, and 29 use previously purified recombinant APE-1 protein that was optimized to the ideal concentration as described here and serves as an appropriate template to optimize the 600 mM imidazole fraction. The final concentration of the inhibitor F327D was 100 μM and the final concentration of the ICM-1 inhibitor was 100 μg/mL. Lanes 1 and 30 show radiolabeled THF DNA without the APE-1 recombinant protein. Lanes 2 and 29 show the APE-1 recombinant protein with the radiolabeled THF DNA (complete cleavage of the THF DNA can be seen). Lanes 3 and 4 show the APE-1 recombinant protein and its THF DNA substrate in the presence of endonuclease activity inhibitors. In lanes 5–8 and 17–20 APE-1 was tested at different dilutions in the presence of DMSO. Lanes 9–12 and 21–24 use the inhibitor F327D and different concentrations of APE-1 recombinant protein. Lanes 13–16 and 25–28 use the inhibitor ICM-1 and different concentrations of APE-1 recombinant protein.

in water), 36 μL H_2O, and 40 μL of corresponding diluted APE-1 recombinant protein (from step 2). Mix gently and thoroughly.

6. Label a 1.5 mL microcentrifuge "A" and add 100 μL of MOPS, 66 μL of glycerol, 62.4 μL of water, and 24 μL of the radiolabeled oligo (see Note 11).
7. Referring to the template created in step 3, add 1.6 μL of DMSO to the microcentrifuge tubes that will be testing the endonuclease activity of the diluted APE-1 recombinant protein in the absence of an inhibitor. Add 1.6 μL of a known APE-1 recombinant protein inhibitor to the microcentrifuge tubes that are to test the endonuclease activity of APE-1 recombinant protein in the presence of an endonuclease inhibitor. The final concentration of the endonuclease activity inhibitor should be 100 μM (see Notes 12 and 17).
8. In a heat block, incubate for 30 min at 37°C.
9. Once the initial 30 min have passed, set a timer to 15 min. Add 7.4 μL of tube "A" to *all* of the microcentrifuge tubes. Begin the timer only after you have added 7.4 μL of "A" to the first microcentrifuge tube (see Note 13).
10. Once 15 min have passed, quench the assay by adding 8 μL of the loading dye beginning with the microcentrifuge tube that received "A" first (see Note 14).
11. The experiment may be stored at 4°C.

3.1.5. Casting of PAGE Gel, Loading of Experiments, Running of Gel, and Imaging of Gel

1. Prepare the gel casting apparatus such that it is dry and clean.
2. In a 100 mL beaker, pour 55 mL of the acrylamide solution prepared in the Subheading 2.5, item 2. Add 360 μL of 10% APS and 36 μL of TEMED to the beaker and mix thoroughly for 20 s.
3. Pour the acrylamide solution into the gel casting apparatus observing that no air bubbles are trapped. Once the acrylamide solution has reached the top, insert the well comb.
4. Allow the acrylamide gel to dry for 30 min.
5. Remove the well comb, observing the integrity of the wells and mount gel on gel running apparatus.
6. Fill the gel running apparatus with 1× TBE so as to cover the wells.
7. Pre-run the gel for 5 min (see Note 15).
8. Load 8–10 μL of loading dye into the first 2–3 wells (see Note 16).
9. Using the template created, load the experiments onto the gel accordingly. Loading dye should be loaded onto the last 2–3 wells.
10. Run the gel at 55 W for 3 h.
11. Remove the gel from the gel casting plate and transfer the gel to a chromatography paper such as a Whatman 3 mm Chr chromatography paper. Cover the gel with saran wrap and transfer the gel to a gel dryer with the saran wrap facing up. Dry for a minimum of 35 min.
12. Place the dried gel (with chromatography paper and saran wrap still in place) into a cassette containing a storage phosphor screen with the gel facing the storage phosphor screen.
13. Allow exposing overnight.
14. Scan the storage phosphor screen using an appropriate imager. Figure 2b displays a representative image.

3.1.6. Recombinant APE-1 Enzymatic Assay

1. Create a template of the compounds that are to be tested for potential APE-1 inhibition. Include a THF DNA and APE-1 control. The THF DNA microcentrifuge tube should initially contain 8.6 μL water. Add 1.6 μL of DMSO to the APE-1 control. Figure 3 shows a schematic of the assay.
2. Make the desired dilutions of the compounds to be tested noting that the final concentration of the assay will be tenfold lower.
3. Label microcentrifuge tubes reflecting the template you have created and aliquot 1.6 μL of the compounds to be tested to the microcentrifuge tubes.

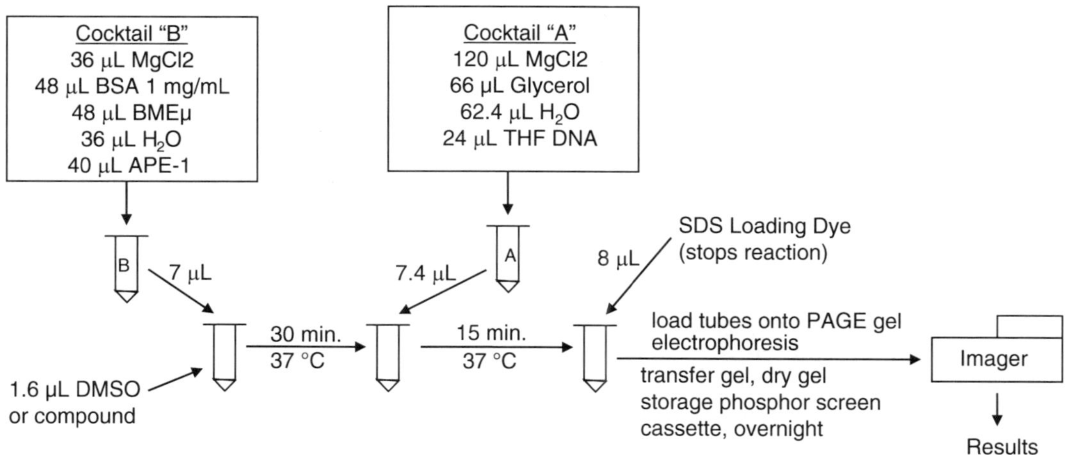

Fig. 3. Schematic of APE-1 endonuclease activity enzymatic assay. Total reaction volume is 16 μL.

4. Set a heat block to 37°C.
5. Label a 1.5 mL microcentrifuge tube "B." Add 36 μL 100 μM $MgCl_2$, 48 μL of 1 mg/mL BSA, 48 μL BME (diluted to 1:50 in water), 36 μL H_2O, and 40 μL of APE-1 recombinant protein. Mix gently and thoroughly.
6. Add 7 μL of "B" and add to every tube containing the compounds to be tested except to the DNA-labeled tube.
7. In a heat block, incubate for 30 min at 37°C.
8. Meanwhile label a tube "A." Add 100 μL of MOPS, 66 μL of glycerol, 62.4 μL of water, and 24 μL of the radiolabeled oligo (see Note 11).
9. Once the initial 30 min have passed, set a timer to 15 min. Add 7.4 μL of tube "A" to *all* of the microcentrifuge tubes. Begin the timer only after you have added 7.4 μL of "A" to the first microcentrifuge tube (see Note 13).
10. Once 15 min have passed, quench the assay by adding 8 μL of the loading dye beginning with the microcentrifuge tube that received "A" first (see Note 14).
11. The experiment may be stored at 4°C.

3.1.7. Loading of Experiments to PAGE Gel, Running of Gel, Drying of Gel, and Imaging of Gel

1. Repeat Subheading 3.1.5, following experiment template. See Fig. 4 for an image of a gel in which a dose response of potential inhibitors was tested.

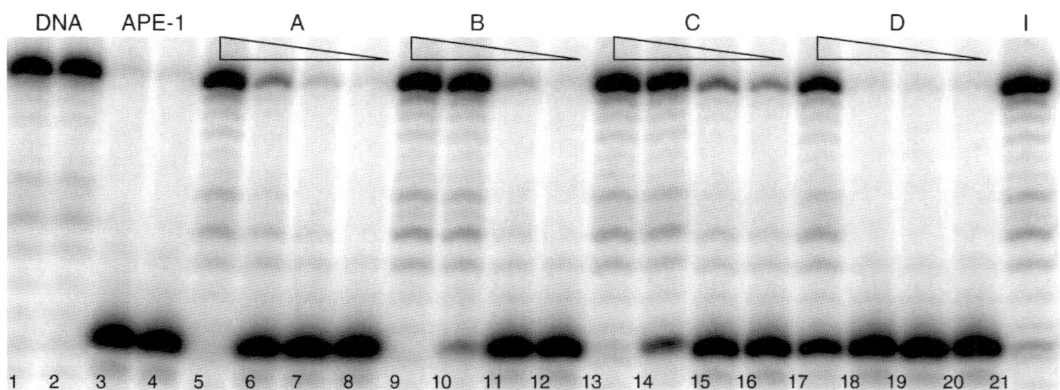

Fig. 4. Representative gel image of dose–response testing of inhibitors of APE-1's endonuclease activity. Lanes 1 and 2: Radiolabeled THF DNA only. Lanes 3 and 4: Radiolabeled THF DNA and APE-1. The following compounds were tested at 100, 33, 11, and 3.7 μM: Lanes 5–8 compound A, Lanes 9–12 compound B, Lanes 13–16 compound C, Lanes 17–20 compound D. Lane 21 contains compound I, an inhibitor of APE-1's endonuclease activity, at a final concentration of 100 μM.

3.2. Fluorescence-Based APE-1 High-Throughput Screening Cleavage Assay

3.2.1. Preparation of Fluorescence-Tagged Oligonucleotides

1. 5′ F-GCCCCCXGGGGACGTACGATATCCCGCTCC3′ and 3′ QCGGGGGCCCCCTGCATGCTATAGGGCGAGG 5′ [where F = Fluorescein, Q = Dabcyl, and X = Tetrahydrofuran (abasic site analog)] should be custom-made for the APE-1 cleavage assay.
2. Anneal the two complementary oligonucleotides by mixing 250 μM of each oligo in a buffer system containing 100 μM Tris–HCl, 50 mM NaCl, and 1 μM EDTA.
3. Heat the mixture to 95°C for 5 min and then allow to cool slowly to room temperature.

3.2.2. Optimization of Fluorescent Oligonucleotide Cleavage Assay

1. Prepare a 5 mL volume of a 1× Fluorescence assay buffer system consisting of 50 mM Tris–HCl, pH 8.0, 1 mM $MgCl_2$, 50 mM NaCl, 2 mM DTT, and 100 mM $MgCl_2$. Using the above-prepared buffer system, perform serial dilutions of the APE-1 recombinant protein. The optimum APE-1 concentration was the minimum concentration that results in statistically significant increase in fluorescence emission compared to the negative control (without APE-1 enzyme).
2. Plates should be incubated at 37°C.
3. Incubation time should be set at 30 min.
4. The Fluorescence readings should be measured every 5 min throughout the 30 min incubation with 495 nm excitation and a 512 nm emission filter.

3.2.3. Fluorescent Oligonucleotide Cleavage Assay for APE-1 Endonuclease Activity (Fig. 5a, b)

1. Incubate a 384-well plate on ice before adding the ingredients of the cleavage assay. This step helps to delay oligonucleotide cleavage by the APE-1 protein.
2. Mark the wells that will be used for testing the compound library.

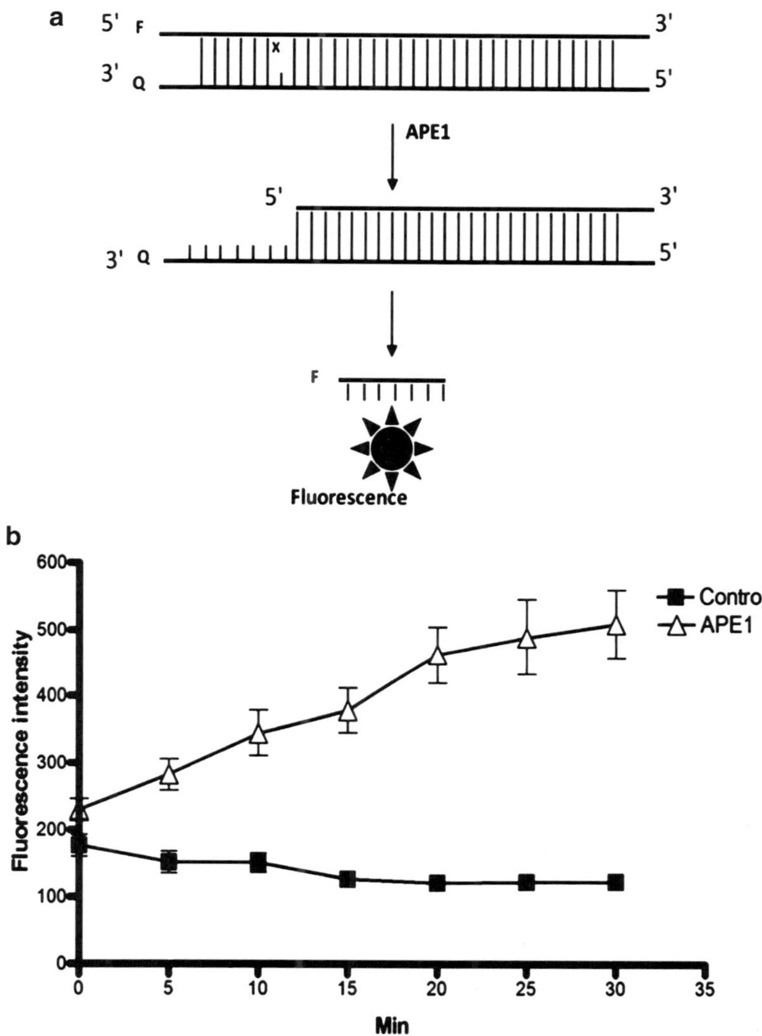

Fig. 5. (**a**) Schematic of the fluorescence-based APE-1 cleavage assay. When the prepared oligonucleotide is cleaved at the abasic site at position 7 from the 5′ end by APE-1, the 6-mer fluorescein-containing product will dissociate from its complement by thermal melting. As a result, the quenching effect of the 3′ dabcyl (which absorbs fluorescein fluorescence when in close proximity) is lost, and APE-1 activity is measured indirectly as an increase in fluorescence signal. (**b**) A typical increase in fluorescence in response to APE-1 endonuclease activity is shown here.

3. Add 14 μL of deionized water to the wells selected for compound testing and shake the plate to ensure that the water is not attached to the well's side wall.
4. Add 4 μL of 10× fluorescence assay buffer and shake the plate briefly.
5. Add 2 μL of recombinant APE-1 protein to the wells selected for compound testing and the wells containing positive control and then shake the plate.

Table 1
Fluorescence oligonucleotide cleavage assay

Time (min)	Negative control			Positive control			CRT0044876		
0	52	55	39	73	54	56	50	54	63
5	54	64	64	75	63	85	46	49	52
10	46	48	47	70	58	78	48	55	45
15	53	49	53	92	72	103	58	50	54
20	50	46	49	115	108	116	56	57	45
25	49	46	50	147	135	148	64	70	62
30	49	40	42	187	174	183	73	66	63

Data shows the fluorescence emission of the DNA substrate in the presence and absence of CRT0044876

6. While the plate is still on ice add 10 μL of the potential inhibitor in the selected wells.
7. Finally add 10 μL of oligonucleotide substrates to the wells containing positive control and negative control and to the wells containing compounds to be tested. The final volume of the reaction mixture is 40 μL.
8. Take a zero time fluorescence intensity reading and incubate the plate at 37°C for a total of 30 min.
9. Take a serial measurement of fluorescence intensity every 5 min for 30 min.
10. A compound that inhibits fluorescence emission after 30 min of incubation is considered as a positive hit (e.g., see Table 1).

3.2.4. IC_{50} Estimation for Potential APE-1 Inhibitors

1. Prepare serial dilutions of the potential inhibitor. For example stock concentrations of 400 μM, 40 μM, 4 μM, 400 nM, and 40 nM will give a final concentration of 100 μM, 10 μM, 1 μM, 100 nM, and 10 nM in the APE-1 cleavage assay.
2. Perform fluorescence cleavage assay at each defined concentration of the potential inhibitor. As before, fluorescence intensity should be measured at 30 min following reaction initiation.
3. Using the initial rate values from the assay, percent activity should be calculated for each sample relative to a negative DMSO-only control.
4. The data should be fitted to a sigmoidal dose–response model using Graphpad Prism3.0 software and IC_{50} values should be determined using the formula: % Activity = 100/(1 + 10 (log[I] − log IC_{50})).

4. Notes

1. Acrylamide is a powerful neurotoxin that is easily inhaled and should be handled with extreme caution. The preparation of the denaturing 20% polyacrylamide will take approximately 20 min. Placing the bottle in hot water will expedite the components of the polyacrylamide mixture to fall into solution. Occasional shaking of the bottle with periodic opening of the cap will also expedite the process.
2. It is helpful to aliquot 1 mL of culture and measure absorbance every 30 min for the first hour and a half. The IPTG should be added at an absorbance reading of 0.600. Because of the logarithmic growth characteristics of bacteria, careful and more frequent absorbance readings (about every 15 min) should be taken near 0.500 absorbance; otherwise the ideal growth phase may be passed.
3. Caution should be taken when using a French press. Have a collection tube on ice ready to collect the lysate. Pass the initial lysate through the French press a second time and collect the lysate in a tube on ice.
4. As different concentrations of imidazole are added to elute the recombinant APE-1 protein, collect each fraction in different tubes. All tubes should be on ice when collecting the fractions.
5. APE-1 recombinant protein is approximately 37 kD. When running an SDS-PAGE gel with the different posteluted fractions, include a protein ladder marker lane.
6. When pouring the resolving gel, leave approximately 25% of the top casting section available for the stacking gel. In essence, pour the resolving gel to approximately 75% the capacity of the gel casting apparatus.
7. Take caution when removing the gel from the plates since the gels are quite fragile. The risk of tearing the gel is reduced if the plates are immersed in running buffer and initially separated at the edges followed by slowly moving away the plates, thereby allowing the running buffer to run into the newly exposed areas of the gel.
8. Protein bands may take up to 4 h to become visible, particularly when the protein concentration is low or the protein bands are narrow.
9. Heating the 1.5 mL reaction tube will denature the T4 PNK, thereby stopping the kinase activity.
10. Centrifuging the radiolabeled oligo duplex will remove free-floating unreacted oligos and materials from the target double-stranded radiolabeled oligo duplex.

11. 100% Glycerol is viscous and will pipette very slowly with a normal 200 μL pipette tip. It is advisable to mix the MOPS, glycerol, and water thoroughly before adding the radiolabeled oligo.
12. The total reaction volume of each microcentrifuge tube/ experiment is 16 μL.
13. It is important that the timer be started only after the first microcentrifuge tube has received 7.4 μL of "A." Since APE-1 is a robust enzyme, it will digest the radiolabeled oligo completely even in the presence of a moderately potent inhibitor if given the time to do so. Thus, keeping an accurate time record of time elapsed since the addition of "A" is essential to the assay.
14. It should take about the same time (±10 s) to add 8 μL of loading dye as it did to add "A." Any duration longer than 1–2 min may greatly affect the results of the assay in reporting the inhibition potential of the compounds tested.
15. Pre-running the gel will equilibrate the ions in the 1× TBE electrophoresis buffer and will provide a constant gel temperature.
16. The first and last 2–3 wells of the PAGE gel will become distorted when electrophoresis is underway. The resulting bands from those wells will not be horizontal. Thus, using loading dye at the ends of the gels increases the possibility of uniform bands resulting at the ends.
17. Comparing the endonuclease activity of the dilutions of the APE-1 recombinant protein in the presence and absence of a known inhibitor of APE-1 endonuclease activity is crucial for the correct determination of APE-1 recombinant protein concentration for subsequent inhibitor testing.

References

1. Zawahir Z, Dayam R, Deng J, Pereira C, Neamati N (2009) Pharamacophore guided discovery of small-molecule human apurinic/apyrimidinic endonuclease 1 inhibitors. J Med Chem 52:20–32
2. Naidu MD, Mason JM, Pica RV, Fung H, Pena LA (2010) Radiation resistance in Glioma cells determined by DNA damage repair activity of APE-1/Ref-1. J Radiat Res 51:393–404
3. Lam W, Park SY, Leung CH, Cheng YC (2006) Apurinic/apyrimidinic endonuclease-1 protein level is associated with the cytotoxicity of L-configuration deoxycytidine analogs (troxacitabine and B-L-2′, 3′-dideoxy2′, 3′-didehydro-5-fluorocytidine) but not D-configuration deoxycytidine analogs (gemcitabine and B-D-arabinofuranosylcytosine). Mol Pharmacol 69:1607
4. Okagbare IP, Soper AS (2010) Polymer-based dense fluidic networks for high throughput screening with ultrasensitive fluorescence detection. Electrophoresis 31:3074–3082
5. Bapat A, Glass LS, Kuo M, Fishel ML, Long EC, Georgiadis MM, Kelley MR (2010) Novel small-molecule inhibitor of Apurinic/Apyrimidinic endonuclease 1 blocks proliferation and reduces viability of glioblastoma cells. J Pharmacol Toxicol Methods 334: 988–998
6. Mohammed MZ, Vyjayanti VN, Laughton CA, Dekker LV, Fischer PM, Wilson DM 3rd, Abbotts R, Shah S, Patel PM, Hickson ID, Madhusudan S (2011) Development and evaluation of human AP endonuclease inhibitors in melanoma and glioma cell lines. Br J Cancer 104:653–66

Chapter 14

An In Vitro Screening to Identify Drug-Resistant Mutations for Target-Directed Chemotherapeutic Agents

Mohammad Azam

Abstract

The discovery of oncogenes and tumor suppressors as a driver of cancer development has triggered the development of target-specific small molecule anticancer compounds. As exemplified by Imatinib (Gleevec), a specific inhibitor of the Chronic Myeloid Leukemia-associated BCR/ABL kinase, these agents promise impressive activity in clinical trials, with low levels of clinical toxicity. However, such therapy is susceptible to the emergence of drug resistance mainly due to amino acid substitutions in the target protein. Defining the spectrum of such mutations is important for patient monitoring and the design of next-generation inhibitors. Using Imatinib and BCR/ABL as a paradigm for a drug–target pair, we reported a retroviral vector-based screening strategy to identify the spectrum of resistance-conferring mutations, which has helped in designing the next-generation BCR/ABL inhibitors such as Nilotinib, Dasatinib, and Ponatinib. Here we provide a detailed methodology for the screen, which can be generally applied to any drug–target pair.

Key words: Protein kinase, Small molecule inhibitors, Tyrosine kinase inhibitors, Acquired drug resistance, Retroviral-screening

1. Introduction

Protein kinases are critical components of signaling networks that catalyze protein phosphorylation and thereby regulate a wide variety of cellular functions including cell proliferation and death, cell adhesion and motility, metabolism, transcription, and differentiation. Spatiotemporal control of phosphorylation is crucial to homeostasis and development, and this control relies on proper regulation of kinases and phosphatases (1–3). An imbalance of phosphorylation/dephosphorylation dynamics can have disastrous consequences for cellular homeostasis often leading to cell transformation, cancer, or a host of other diseases (4). More than 400 human diseases have been linked, directly or indirectly, to protein kinases (5). Therefore, the ability to modulate kinase activity represents an attractive

Yi Zheng (ed.), *Rational Drug Design: Methods and Protocols*, Methods in Molecular Biology, vol. 928,
DOI 10.1007/978-1-62703-008-3_14, © Springer Science+Business Media New York 2012

therapeutic intervention for the treatment of various human illnesses (6–8).

Novel discoveries in cancer biology have provided the opportunity to design target-specific anticancer agents and fostered rapid advancements in drug development. The current focus is the design of molecules with high selectivity against specific proteins in malignant cells to insure a high therapeutic index with minimal side effects. The unprecedented success of the BCR/ABL tyrosine kinase inhibitor Imatinib (Gleevec) in the treatment of chronic myeloid leukemia (CML) has inspired great expectations for this approach (9, 10). Complete hematologic responses to Imatinib are seen in >95% of CML patients and a major cytogenetic response in >60% of patients treated in the chronic phase of the disease (7). However, highly specific protein inhibition brings with it a critical problem: protein targets develop escape mutations leading to drug resistance. In fact, virtually all patients with advanced stages of CML ultimately manifest Imatinib resistance (11–13), and it was expected that other protein targets would evolve drug-resistant forms as well in response to therapy. Indeed, pharmacological inhibition of EGFR, c-KIT receptor, and PDGFR-alpha developed resistance to therapy by acquiring point mutations that directly or indirectly affect drug binding, viz., drug resistance. The identification of these mutant forms is essential for the design of more robust next-generation therapies, and may ultimately lead to molecular cocktails designed to circumvent resistance.

In previous work, we have reported the results of our screen involving random mutagenesis of BCR/ABL to reveal the spectrum of mutations conferring resistance to BCR/ABL (14, 15) inhibitors such as Imatinib, PD166326 (15), and AP4163 (16). The results not only identified the mutants critical for clinical disease relapse but also shed light on the structural regulatory mechanisms of kinases (14, 17). Here we provide additional methodologic detail to enable a broader application of this screening strategy to additional drug–target pairs.

2. Materials

2.1. Plasmids

1. pEYK3.1 or POP-PURO-Ires-GFP retroviral vectors for target gene expression.
2. pCL-ECO ecotropic helper plasmid for the generation of retroviruses.

2.2. E. coli Strains, Reagents for Microbial Culture, and Selection for Transformants

1. TOP10 Electrocompetent cells.
2. TOP10 Chemical competent cells.
3. X-GAL.

4. LB media.
5. LB agar.
6. XL-1 red competent cells (Agilent technologies, cat. # 200129).

2.3. Cell Lines

1. 293T cells (American Type Culture Collection, cat. # CRL11268).
2. BAF3 cell (DSMZ# ACC 300).

2.4. Cell Culture Media

1. DMEM.
2. RPMI.
3. Heat-inactivated FBS.
4. Penicillin/streptomycin solution 100×.
5. 70% (vol/vol) ethanol.
6. PBS without Ca/Mg.
7. GlutaMAX.
8. 2-Mercaptoethanol.
9. WEHI conditioned media.

2.5. Reagents for DNA Transfection and Viral Transduction

1. Protamine sulfate.
2. FuGENE 6 transfection reagent.

2.6. Antibiotics

1. Zeocin.
2. Puromycin.
3. Kanamycin.

2.7. Reagents for DNA Manipulation

1. PCRXL-TOPO Kit.
2. Rapid DNA ligation kit.
3. EXPAND Long template PCR.
4. QUICKCHANGE XL mutagenesis kit.

2.8. Reagents for Protein Analysis

1. Protease inhibitor cocktail from Roche.
2. SuperSignal West Pico Chemiluminescent Substrate.
3. Phosphatase Inhibitor Cocktail 1 (Sigma; cat. # P2850).
4. Phosphatase Inhibitor Cocktail 1 (Sigma; cat. # P5726).
5. Carnation Nonfat dry milk.

2.9. Instruments for Cell Culture, DNA, and Protein Analysis

1. PCR Thermal cycler.
2. DNA gel electrophoresis system.
3. Protein gel electrophoresis system.
4. Protein gel transfer apparatus.

5. Microcentrifuge.
6. 37 C incubator shaker for bacterial culture.
7. Inverted tissue culture microscope with phase contrast (4, 10, 20, 40 objectives).
8. Stereomicroscope.
9. Biosafety cabinet with aspirator for tissue culture.
10. CO2 incubator, 37°C, humidity.
11. Tissue culture centrifuge.
12. Tissue culture dish, 35, 100, and 150 mm.
13. Tissue culture plates, 6 well, 12 well, and 96 well.
14. Conical tubes, 15 and 50 ml.
15. Glass Pasteur pipettes, 9 in., sterilized using autoclave.
16. Cryovials, 2.0 ml.
17. Plastic disposable transfer pipettes, 1, 5, 10, 25, and 50 ml.
18. Disposable sterile filter system, 0.22 m, 500 ml.
19. Disposable syringes, 10, 5, and 1 ml.
20. Hypodermic needle, 27–30 G.
21. Acrodisc filter, 0.45 mM, low protein binding.
22. Acrodisc filter, 0.2 mM, DMSO safe.
23. Sterile Petri dishes.
24. Coulter counter (Beckman Coulter) or hemocytometer.

3. Methods

To identify a wide spectrum of drug resistance conferring mutations, we generated a high-complexity library of mutagenized BCR–ABL cDNA in a retroviral vector and introduced this into cells by retroviral transduction. We then selected for surviving drug-resistant clones, recovered plasmid DNA or PCR amplicons, and analyzed their sequence for the presence of mutations. To verify that the observed mutations were the basis of drug resistance, we regenerated each mutation separately by site-directed mutagenesis (SDM) of native BCR–ABL, and reintroduced them into cells. Briefly, the methodology is as described earlier (Fig. 1) (14, 18); clone the target cDNA into a retroviral vector. Propagate the vector in bacteria deficient in DNA repair mechanisms, creating an exhaustive library of mutations in the target gene. Transfect/infect drug-sensitive cells with the mutated vector and disperse in soft agar in the presence of drug. Isolate resistant colonies, recover the target cDNA, and sequence to identify mutations. The resistant phenotype

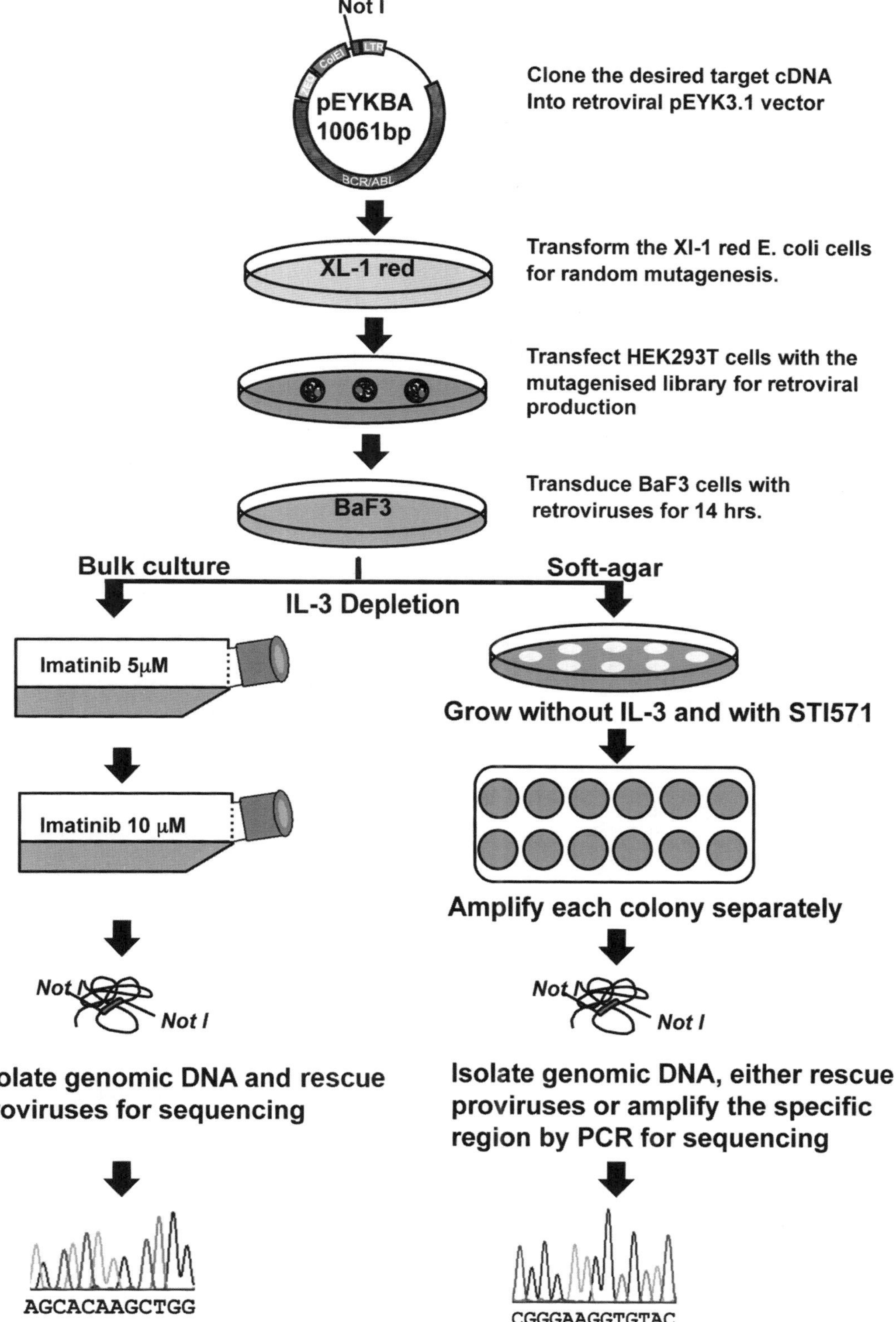

Fig. 1. A general scheme for drug-resistant screening of a drug–target pair.

of the mutations is then confirmed recreating the mutation in the native cDNA by SDM. Cells are transfected/infected and grown in the presence of drug. Resistance is measured by proliferation assays and/or immunoblotting. Additionally, to gain insight about the structural consequences on drug binding and resistance, mutations are mapped on a model of the protein crystal structure to perform *in silico* analysis.

3.1. Retroviral Construct and Random Mutagenesis

1. Clone the desired target gene in either pEYK3.1 (19) or pOP-puro-Ires GFP vector (17) (see Note 1).
2. Transform the recombinant clone to XL-1 red *E. coli* cells. Thaw 100 μl XL-1 Red competent *E-coli* (Stratagene) in a polypropylene tube on ice. Gently mix in 10–50 ng plasmid DNA. Incubate on ice for 30 min, gently swirling every 2 min. Immerse the tube in a 42°C bath for 45 s and immediately incubate on ice for 2 min. Add 1 ml SOC medium and incubate at 37°C shaking at 225–250 rpm for 90 min.
3. Plate the transformation mix on LB-agar plate containing zeocin (25 ug/ml) and incubate at 37°C for 36–48 h (see Note 2).
4. Collect the colonies by scraping the plates with a sterile plate scraper. Isolate plasmid DNA using the Qiagen midi-prep Kit. At this stage, the heterogeneity of mutations in the library can be roughly assessed by restriction digestion with a frequent cutter such as *Sau3A1* or *Taq1* or by DNA sequencing.

3.2. Production of Retroviral Supernatants

1. One day before transfection, plate 2×10^6 HEK 293 T cells onto six 100 mm dishes in DMEM containing 10% FCS, pen/strep, and 2 mM L-glutamine.
2. Replace the medium the next day, and transfect the cells with mutagenized library of target gene. Transfect each 100 mm plate with 10 μg of DNA (mix 5 μg of mutagenized retroviral library with 5 μg of retroviral packaging construct pCL-eco[8] and then add 30 μl of FuGENE6).
3. Pour the transformation mix dropwise on the HEK293T cells. Incubate the plate overnight at 37 C.
4. Change media the next day, and carefully add 10 ml of medium over the cells as they are loosely attached to the surface.
5. Collect the media after 48 h containing retroviruses. Filter the viral supernatant through a 0.45 μm Acrodisc filter to remove cell debris and particulate matters. The retroviruses can be snap frozen with liquid nitrogen and stored at −80°C or used immediately for transduction experiments.

3.3. Retroviral Transduction

1. Grow BaF3 cells in R10 media (RPMI with 10% FCS, 100 U Penicillin/100 microgram Streptomycin per ml, 2 mM L-glutamine, and 10% WEHI-3B conditioned medium (as a

source of IL-3). The BaF3 cell line is a murine pro B cell line that is dependent on Interleukin-3 (IL-3) for growth. Ectopic expression of constitutively activated tyrosine kinases such as BCR–ABL, EGFR, SRC, PDGFR, and JAK2-V617F renders the BaF3 cells IL-3 independent[10].

2. Transfer 1×10^6 BaF3 cells in each well of 6-well tissue culture dish.
3. Add 1 ml of viral supernatant having a titer of 10^5–10^6 infectious units/ml (see Note 3).
4. Add 3 μl of polyberene (8 mg/ml) and 1 ml of R10 media with WEHI.
5. Centrifuge the plate for 90 min at 1,500×g in a Sorvall RT 6000 table centrifuge.
6. Transfer the plates to the incubator for 14–16 h at 37°C with 5% CO_2 (see Note 4).

3.4. Screening for Resistant Clones

1. To select for drug-resistant clones, plate the virally transduced cells in soft agar containing varying concentrations of drugs. Mix 8×10^6 cells with 28.8 ml RPMI, 9.6 ml FCS, 9.6 ml of 1.2% Bacto-agar (made in PBS, autoclaved, and cooled to 42°C), and supplemented with varying concentrations of drugs (1–20 μM).
2. Plate 3 ml of cell-mix per well into 6-well plates and incubate at 37°C, 5% CO_2, for 14–24 days.
3. Pick the single colonies and expand separately in 3 ml R10 media in the presence of drugs.
4. Collect the cells after reaching to confluence. Isolate the genomic DNA using the Qiagen DNeasy Kit.
5. Target DNA can be sequenced either by rescuing the whole gene followed by subcloning or by sequencing PCR amplicons of desired region of the target gene (see Note 5).

3.5. Provirus Rescue from Genomic DNA

1. Digest 10 μg of genomic DNA with *NotI* to release the provirus.
2. Ligate the digested DNA (100 μl of reaction volume per 1 μg of genomic DNA) using T4 DNA ligase from Roche.
3. Extract the DNA using phenol/chloroform extraction and ethanol precipitation.
4. Resuspend the precipitated DNA in 10 ml of water. Transform the TOP10 competent cells (Invitrogen) with ligated DNA.
5. Plate the transformation mix on LB-agar/zeocin plates.
6. Pick the colonies, and isolate the plasmid DNA using Qiagen mini-prep kit for sequencing (see Note 6).

3.6. Site-Directed Mutagenesis

1. To specifically mutate a desired residue we performed SDM on our plasmids using the Quickchange Mutagenesis Kit from Stratagene and oligonucleotides that were designed to create the point mutations found in our screen. Confirm the mutations by DNA sequencing (see Note 7).

3.7. Cell Viability Assay to Determine IC50

1. Plate 10^4 BaF3 cells expressing mutant genes into each well of a 96-well plate in RPMI/10% FCS.
2. Add the drugs to the media in increasing concentrations (final concentration: 0, 1, 3, 5, 10, and 20 μM) across the plate.
3. Incubate the plate for 48–60 h at 37°C, 5% CO_2.
4. Assess the cell viability by adding 10 ml of WST-1 reagent (Roche).
5. Read the plate with an ELISA plate reader at 450 nm (see Note 8).

4. Notes

1. We recommend using either pEYK3.1 (19) or pOP-puro-Ires-GFP (17) vectors for screening because they have single LTR with an engineered Not I site, which facilitates the proviral rescue from genomic DNA. Additionally, single LTR-based vector can be used directly for SDM, as conventional retroviral vectors having two LTRs are not good substrate for SDM reaction because they tend to loop out the inserts during PCR amplification.
2. Because XL-1 red cells are growing cells, colonies start appearing after 24 h. We recommend harvesting the cells as soon as colonies start appearing (typically 24–36 h but in some cases it may take a little longer depending on the type of genes) followed by plasmid miniprep. Use this purified DNA to transform the normal XL-1 blue *E. coli*-competent cells to amplify the plasmid library.
3. We have employed low viral transduction efficiency (e.g., 20–30%) in order to avoid inducing a drug-resistant phenotype due to multiple integration of provirus. Therefore we recommend not to use high-titer viral supernatant or overexpression of the protein to a specific level.
4. During our screen, we found that it is important to avoid bulk culture conditions in which cells harboring different mutations are pooled together and allowed to expand in liquid culture, since this can lead to clonal dominance of a few highly drug-resistant variants. When we initially selected for imatinib

resistance of mutagenized BCR/ABL in bulk liquid culture, we found only 4 mutant forms that were represented multiple times within the first 100 isolates we sequenced. In contrast, we identified over 100 mutants when we selected for outgrowth of clones in the soft agar colony-forming assay. Some slow-growing clones with more modest degrees of drug resistance could not have been recovered using bulk culture. We thus recommend to avoid any growth or selection in bulk, within either the bacterial or mammalian cell cultures. For that reason, we attempted to minimize any incubation periods where cells are pooled together. After retroviral infection, we maintained the cells in bulk culture prior to selection for only 14–16 h, after which we performed combined IL-3 withdrawal and drug selection in soft agar. This provided for tight selection of individual clones and prevented clonal dominance of liquid cultures. If the cells used for drug screening do not form colonies in soft agar they can be selected in a similar clonal manner by limiting dilution in 96-well plates.

5. We recommend using PCR amplification of the desired region in a given gene but in some instances we failed to amplify the desired target due to complexities incurred by genomic DNAs. In such cases we performed the proviral rescue to clone whole plasmid again.
6. Proviruses can be rescued by Not I digestion followed by ligation/transformation. However, in some cases transgenes have an internal Not I site, e.g., SRC kinase; in this situation we used Cre recombinase mediate excision of proviruses by incubating the genomic DNA with recombinant Cre for 30 min at 37 C followed by transformation plating on zeocin plate. In pEYK3.1 and pOP-puro-Ires-GFP vectors, we have engineered Cre recombination site flanking the LTR. Because of low efficiency of Cre recombination as compared to Not I digestion, we do not recommend to use Cre-mediated provirus rescue as a first choice but it can be used in the cases when the transgenes are harboring internal Not I sites.
7. We recommend using freshly prepared plasmids as a substrate for SDM reaction. Older plasmid preparation or nicked plasmids give lots of background and often fail to produce the desired mutant clones.
8. All assays should be performed in quadruplicate and readings are averaged and plotted against drug concentration as a best-fit sigmoidal curve using a nonlinear curve-fitting algorithm (Origin 7.0, Origin Lab, Northampton, MA). The drug concentration resulting in 50% cell viability scored as the Cellular IC50.

Acknowledgment

This work was supported by grants from V Foundation.

References

1. Azam M, Daley GQ (2006) Anticipating clinical resistance to target-directed agents: the BCR-ABL paradigm. Mol Diagn Ther 10:67–76
2. Hanahan D, Weinberg RA (2000) The hallmarks of cancer. Cell 100:57–70
3. Blume-Jensen P, Hunter T (2001) Oncogenic kinase signalling. Nature 411:355–65
4. Huse M, Kuriyan J (2002) The conformational plasticity of protein kinases. Cell 109:275–82
5. Melnikova I, Golden J (2004) Targeting protein kinases. Nat Rev Drug Discov 3:993–4
6. Cohen P (2002) Protein kinases–the major drug targets of the twenty-first century? Nat Rev Drug Discov 1:309–15
7. Dancey J, Sausville EA (2003) Issues and progress with protein kinase inhibitors for cancer treatment. Nat Rev Drug Discov 2:296–313
8. Noble ME, Endicott JA, Johnson LN (2004) Protein kinase inhibitors: insights into drug design from structure. Science 303:1800–5
9. Druker BJ, Sawyers CL, Kantarjian H, Resta DJ, Reese SF, Ford JM, Capdeville R, Talpaz M (2001) Activity of a specific inhibitor of the BCR-ABL tyrosine kinase in the blast crisis of chronic myeloid leukemia and acute lymphoblastic leukemia with the Philadelphia chromosome. N Engl J Med 344:1038–42
10. Druker BJ, Tamura S, Buchdunger E, Ohno S, Segal GM, Fanning S, Zimmermann J, Lydon NB (1996) Effects of a selective inhibitor of the Abl tyrosine kinase on the growth of Bcr-Abl positive cells. Nat Med 2:561–6
11. Gorre ME, Mohammed M, Ellwood K, Hsu N, Paquette R, Rao PN, Sawyers CL (2001) Clinical resistance to STI-571 cancer therapy caused by BCR-ABL gene mutation or amplification. Science 293:876–80
12. Shah NP, Nicoll JM, Nagar B, Gorre ME, Paquette RL, Kuriyan J, Sawyers CL (2002) Multiple BCR-ABL kinase domain mutations confer polyclonal resistance to the tyrosine kinase inhibitor imatinib (STI571) in chronic phase and blast crisis chronic myeloid leukemia. Cancer Cell 2:117–25
13. Branford S, Rudzki Z, Walsh S, Grigg A, Arthur C, Taylor K, Herrmann R, Lynch KP, Hughes TP (2002) High frequency of point mutations clustered within the adenosine triphosphate-binding region of BCR/ABL in patients with chronic myeloid leukemia or Ph-positive acute lymphoblastic leukemia who develop imatinib (STI571) resistance. Blood 99:3472–5
14. Azam M, Latek RR, Daley GQ (2003) Mechanisms of autoinhibition and STI-571/imatinib resistance revealed by mutagenesis of BCR-ABL. Cell 112:831–43
15. Azam M, Nardi V, Shakespeare WC, Metcalf CA 3rd, Bohacek RS, Wang Y, Sundaramoorthi R, Sliz P, Veach DR, Bornmann WG, Clarkson B, Dalgarno DC, Sawyer TK, Daley GQ (2006) Activity of dual SRC-ABL inhibitors highlights the role of BCR/ABL kinase dynamics in drug resistance. Proc Natl Acad Sci USA 103: 9244–9
16. Azam M, Powers JT, Einhorn W, Huang WS, Shakespeare WC, Zhu X, Dalgarno D, Clackson T, Sawyer TK, Daley GQ (2010) AP24163 inhibits the gatekeeper mutant of BCR-ABL and suppresses in vitro resistance. Chem Biol Drug Des 75:223–7
17. Azam M, Seeliger MA, Gray NS, Kuriyan J, Daley GQ (2008) Activation of tyrosine kinases by mutation of the gatekeeper threonine. Nat Struct Mol Biol 15:1109–18
18. Azam M, Raz T, Nardi V, Opitz SL, Daley GQ (2003) A screen to identify drug resistant variants to target-directed anti-cancer agents. Biol Proced Online 5:204–10
19. Koh EY, Chen T, Daley GQ (2002) Novel retroviral vectors to facilitate expression screens in mammalian cells. Nucleic Acids Res 30:e142

Chapter 15

Utilizing AntagomiR (Antisense microRNA) to Knock Down microRNA in Murine Bone Marrow Cells

Chinavenmeni S. Velu and H. Leighton Grimes

Abstract

MicroRNAs (miRNAs) are highly conserved small RNAs which regulate gene expression primarily through base pairing to the 3′ untranslated region of target messenger RNA (mRNA), leading to mRNA degradation or translation inhibition depending on the complementarity between the miRNA and target mRNA. Single miRNA regulates multiple target mRNA. miRNAs have been shown to regulate gene expression in the hematopoietic stem cells, as well as at key decision points for various lineages. However, aberrant expression of miRNAs has been documented in cancer and disease models. Rigorous dissection of miRNA pathways and biology requires facile loss of function modeling. This chapter describes detailed protocol for knockdown miRNA-21 which is involved in myelopoiesis using antagomiRs in primary murine bone marrow stem/progenitor cells.

Key words: microRNA, Antagomir, Murine bone marrow cells, Drug delivery

1. Introduction

MicroRNAs (miRNAs) are endogenous regulatory small RNA molecules of about 17–22 nucleotides in length and have been described recently as modulators of gene activity. miRNAs are expressed as primary miRNA transcripts which are processed to precursor miRNA and then acted upon by DROSHA–DGCR protein complexes to form mature miRNAs (1). miRNA binds to target messenger RNA resulting in posttranscriptional or post-translational repression (2). Almost 700 miRNAs have been identified in humans regulating a predicted one-third of protein-coding genes. Moreover, one miRNA is believed to target multiple mRNAs; similarly one mRNA may be targeted by multiple miRNA highlighting the complex nature of gene expression regulation (3). However, the role of miRNA in physiological and pathological processes is still emerging with mounting evidence to suggest

Yi Zheng (ed.), *Rational Drug Design: Methods and Protocols*, Methods in Molecular Biology, vol. 928, DOI 10.1007/978-1-62703-008-3_15, © Springer Science+Business Media New York 2012

that miRNAs are involved in a remarkable spectrum of biological pathways (4, 5). Different sets of miRNAs are expressed in different cell types and tissues which likely reflect distinct and highly specific regulatory processes. Aberrant expression of miRNAs has been implicated in numerous disease states (6, 7).

Pluripotent hematopoietic stem cells (HSCs) express a variety of lineage-specific miRNAs which may enforce lineage commitment by limiting the expression of proteins from alternative lineages (8). Understanding the role of miRNAs in hematopoietic stem/progenitor cells (HSPCs) requires strategies utilizing overexpression and/or efficient loss-of-function. We have accomplished this through a variety of in vitro assay systems such as colony-forming cell-unit (CFU) assays. The clonogenic hematopoietic (CFU) assay detects progenitor cells committed to a specific lineage by plating HSPCs into semisolid media such as methylcellulose together with cytokines and hematopoietic growth factors to stimulate clonal proliferation into myeloid-restricted lineages that include granulocyte, monocyte, and erythrocyte cells. Overexpression studies are performed by using vectors or chemically engineered oligonucleotides that "mimic" mature miRNA and have the potential to treat or cure cancers (9, 10). In loss-of-function studies, vectors for miRNA knockdown (11, 12) or modified antisense oligonucleotides for mature miRNA known as antagomiRs have been used (13). In this chapter, we describe an in vitro and in vivo assay model to reduce miRNA-21 function in murine bone marrow cells.

2. Materials

2.1. Lineage Depletion and Colony-Forming Unit Assay

1. Lineage Cell Depletion Kit for mouse, AutoMACS™ Separator, and magnetic column.
2. Running Buffer: PBS (phosphate buffered saline) pH 7.2, 0.5% BSA (bovine serum albumin), 2 mM EDTA.
3. Rinsing Buffer: PBS pH 7.2, 2 mM EDTA.
4. 70% alcohol.
5. Cytokines: mSCF; 10 μg, mIL-3, 10 μg.
6. StemSpan and MethoCultGF.
7. 45 μM cell strainer.
8. 0.45 μM syringe filter.
9. 6 cm tissue culture grid plate.
10. Non-treated tissue culture plates: BD Multi-well cell culture plate 24-well, 6-well.
11. AntagomiRs.

2.2. Immunoblotting

1. Complete Lysis M Buffer.
2. Protease Buffer.
3. 10% Separating gel; 0.375 M Tris–HCl, pH 8.8, 10% Acrylamide, 0.1% SDS, 0.1% TEMED, 0.1% APS.
4. 5% Stacking gel; 0.125 M Tris–HCl, pH 6.8, 5% Acrylamide, 0.1% SDS, 0.1% TEMED, 0.1% APS.
5. SDS-Polyacrylamide Gel Electrophoresis (SDS-PAGE) running buffer: 25 mM Tris base, 250 mM glycine, 0.1% SDS.
6. Transfer buffer: 25 mM Tris base, 192 mM Glycine, 20% methanol.
7. Polyvinylidene difluoride (PVDF) membrane (Immobilon-P IVPH00010).
8. Phosphate-buffered saline with Tween (PBST) 1× PBS, 0.05% Tween 20.
9. Blocking buffer: 5% nonfat dry milk in PBST.
10. Enhance chemiluminescent (ECL) plus western blotting detection system.
11. Stripping buffer: 0.1 M Glycine, 125 mM HCl.

2.3. Surgery and Implanting the Pumps

1. Alzet Osmotic Pumps.
2. 1 mL Syringe with pump loading needle.
3. 1 mL Syringe with 27G needle for Avertin delivery and Brupenex.
4. Wound Stapler and Wound Clips.
5. Blunt-tip dissection scissors.
6. Sharp Scissors.
7. Two pairs of forceps (45°).
8. Shaver.
9. 0.9% Sterile saline.
10. Isoflurane.
11. Avertin (anesthetic); prepare a stock solution by dissolving 25 g of avertin in 15.5 mL of *tert*-Amyl alcohol and the working solution (1.2%) was prepared by diluting 0.5 mL of avertin stock with 39.5 mL 0.9% saline (working solution). Inject ~250 μL/mouse.
12. Brupenex (analgesic).
13. 6–8-week-old C57/B6 mice.

3. Methods

The design of a specific antagomiR for an miRNA loss-of-function phenotype requires the knowledge of the mature miRNA sequence. The most common Web site that offers miRNA sequences is miRBase (www.mirbase.org) (14) (see Note 1).

3.1. Designing antagomiR

Example of antagomiR template design

Example—miRNA-21

Target mature miRNA sequence

5′-UAG CUU AUC AGA CUG AUG UUG A-3′

Antisense mature miRNA sequence

3′-AUC GAA UAG UCU GAC UAC AAC U-5′

AntagomiR

5′-U CAA CAU CAG UCU GAU AAG CUA-3′

Mutant AntagomiR (control)

5′-U CAA C**U**U CAG UC**A** GA**A** AAG **G**UA-3′

3.2. Synthesis of antagomiR

AntagomiRs are synthetic 2-*O*-methyl RNA oligos of about 21–23 nucleotides which fully complement the miRNAs and effectively compete with miRNA target mRNAs with a stronger binding to the miRNA-associated gene silencing complexes (miRNA-RISCs). AntagomiRs were synthesized and purchased from Dharmacon. AntagomiRs were synthesized with 2′-OMe modified bases (2′-hydroxyl of the ribose is replaced with a methoxy group), phosphorothioate (phosphodiester linkages are changed to phosphorothioates) on the first two and last four bases, and an addition of cholesterol motif at 3′ end through a hydroxyprolinol modified linkage. The addition of 2′-OMe and phosphorothioate modifications improve the bio-stability whereas cholesterol conjugation enhances distribution and cell permeation of the antagomiRs (15). AntagomiR oligonucleotides are deprotected, desalted, and purified by high-performance liquid chromatography.

3.3. Preparing the antagomiR

1. Centrifuge the tube briefly to pellet.
2. Reconstitute the antagomiR (powder) with nuclease-free sterile water at 3× dosing concentration.
3. Mix the suspension at room temperature until the antagomiR fully goes into the solution (see Note 2).
4. Filter the antagomiR through a 0.45 μM syringe filter.
5. To calculate the concentration, dilute a small aliquot of antagomiR to 1/250 to 1/500 dilution and determine the Optical Density (OD) using Nanodrop.

Example: For calculating mM concentration and mg/mL using extinction coefficient and molecular weight:

$$(\text{OD} \times \text{dilution factor}) / \text{extinction coefficient} = \text{mM concentration}$$

$$(\text{mM concentration} / 1{,}000) \times \text{MW} = \text{mg} / \text{mL}$$

If you know the molar concentration, it is easy to prepare a stock solution, e.g., if you get 250 nM of antagomiRs, dissolving in 500 μL of water gives you a 500 μM stock solution.

6. Keep the antagomiR in small aliquots to avoid freeze and thaw and for long-term storage at −20°C.

3.4. *In Vitro* Assay

3.4.1. Isolating Mouse Bone Marrow Cells

1. Flush the bone marrow cells from both femur and tibias with Medium 199 using a syringe and a 23 gauge needle.
2. Disaggregate the cells by gentle pipetting and pass the cells through 70 μm Nylon cell strainer to remove cell clumps and bone pieces.
3. Wash the cells by adding running buffer and centrifuge at 300 × *g* for 10 min at 4–8°C.
4. Pipette off the supernatant completely.
5. Resuspend the cell pellet in running buffer and determine the cell number.
6. Label the cells with Lineage cells depletion kit according to the manufacturer's instructions.
7. Centrifuge at 300 × *g* for 10 min and resuspend cell pellet in 40 μL of buffer per 10^7 total cells.
8. Add 10 μL of Biotin–Antibody Cocktail per 10^7 total cells.
9. Mix well and incubate on ice for 10 min.
10. Add 30 μL of buffer per 10^7 total cells.
11. Add 20 μL of Anti-biotin MicroBeads per 10^7 total cells.
12. Mix well and incubate on ice for additional 15 min.
13. Wash cells by adding 1–2 mL of buffer per 10^7 cells and centrifuge at 300 × *g* for 10 min.
14. Resuspend up to 10^8 cells in 500 μL of buffer.
15. Purify the lineage negative cells using AUTOMACS magnetic sorter or magnetic column.
16. Collect the Linage negative cells and wash the cells with PBS with 0.5% FBS and count the cells.

3.4.2. Inhibiting microRNA Activity

1. For liquid culture, the cells are maintained at a density of 5×10^5 cell/mL in serum-free medium, 25 ng/mL mSCF, 20 ng/mL IL6, and 10 ng/mL IL3, and incubate at 37°C, 5% CO_2, 100% humidity in a cell culture incubator.

2. For antagomiR treatment, add 1 μL of 100 μM stock solution to 1 mL of culture medium to get a final concentration of 100 nM in a 24-well non-tissue culture plate.
3. Mix well and incubate the cells in the cell culture incubator.

Assay for Target Gene Activity: Taqman Expression

For mature miRNA expression analysis, total cellular RNA is extracted from antagomiR-treated cells and subjected to specific miR-qRT-PCR. As shown in Fig. 1a, murine bone marrow treated with antagomiR-21 shows a reduced level of mature miR-21 compared to the cell treated with specific mutant control.

1. To check endogenous miRNA expression, isolate total cellular RNA from the treated cells using TRI reagent at various time points (24, 48, 72 h).
2. Reverse transcribe the RNA using TaqMan® MicroRNA Reverse Transcription Kit, according to the manufacturer's protocol using the commercially available assay for specific mature miRNA.
3. Perform the amplification reaction using specific miRNA taqman probe on ABI Prism 7900 amplification system.
4. Normalization and analysis of miRNA expression can be calculated using the comparative 2ΔΔ Ct method relative to snRNA-U6.

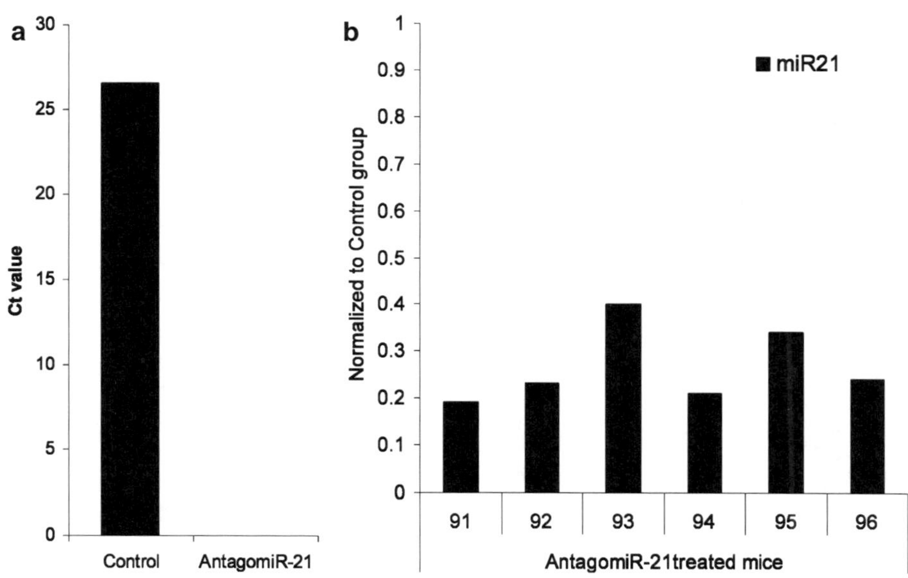

Fig. 1. Silencing miR-21 using antagomiR-21 (**a**) in liquid culture (**b**) in vivo in mice. For in vitro assay, wild-type Lin-bone marrow cells were treated with either control or antagomiR-21 at a concentration of 100 nM for 72 h. For in vivo assay, the mice were implanted with either control or antagomiR-21 and the expression of mature miR-21 is normalized to control mice (set as 1). The results shown here are from individual mice.

Immunoblotting Analysis of miRNA Target Proteins

Effective miRNA silencing should result in upregulation of miRNA target proteins. AntagomiR functionality can be analyzed by using an immunoblotting approach. As shown in Fig. 2 mouse bone marrow cells treated with antagomiR-21 show an increased level of miR-21 target protein Pten (phosphatase and tensin homology) as compared to control-treated cells.

1. Wash the cells twice with cold 1× PBS, add cold Complete M Lysis buffer, and isolate protein according to the manufacturer's instruction.
2. Separate 40 μg of protein by electrophoresis on 10% SDS-PAGE under reducing and denaturing conditions (100 V and 40 mA for 2 h).
3. After gel electrophoresis, proteins are transferred to a PVDF membrane (15 V for 30 min).
4. Incubate the membrane in blocking buffer for 1 h at room temperature, followed by washing three times with 1× PBST.
5. Add primary antibody in blocking buffer and incubate the membrane overnight at 4°C.
6. Wash the membrane three times with 1× PBST and incubate with horseradish peroxidase (HRP)-conjugated secondary antibody for 1 h at room temperature.
7. Wash the membrane three times with 1× PBST.
8. Use ECL plus system to visualize the immune complexes.
9. For reprobing, incubate the membrane at room temperature for 1 h in stripping buffer followed by intensive washing with 1× PBST.
10. The membrane is then ready to be used for probing for beta-Actin as loading control.

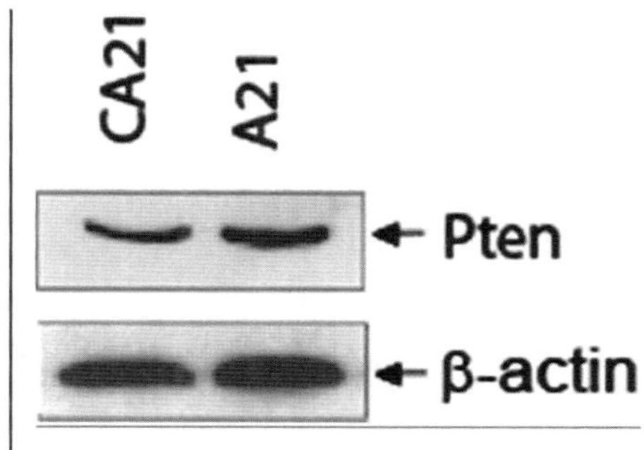

Fig. 2. Effect of miR-21 knockdown on its target protein Pten in wild-type mouse bone marrow cells. Cells were treated with either control or antagomiR-21 in liquid culture, harvested with protein lysate after 72 h, and analyzed with respective antibodies.

Assay for Biological Activity: Colony-Forming Unit Assays

Phenotypic analysis of proliferation, survival, and differentiation of treated cells can be easily monitored by several standard techniques. As we have published (16) and shown in Fig. 3a, mouse bone marrow cells treated with antagomiR-21 show a decrease in total colony numbers with a specific increase in the monocytic colonies (CFU-M).

1. For CFU assay, take 5×10^3 Lin-mouse bone marrow cells in 100 μL of StemSpan in a 5 mL tube and add 1 μL of 100 μM antagomiR.
2. Incubate the cells with antagomiR on ice for 30 min.
3. Add 1 mL of room-temperature MethoCultGF M3535 to the 5 mL tube (see Note 3).
4. Mix the cells vigorously with a 2 mL pipette and keep it in room temperature for 5 min to get rid of air bubbles.

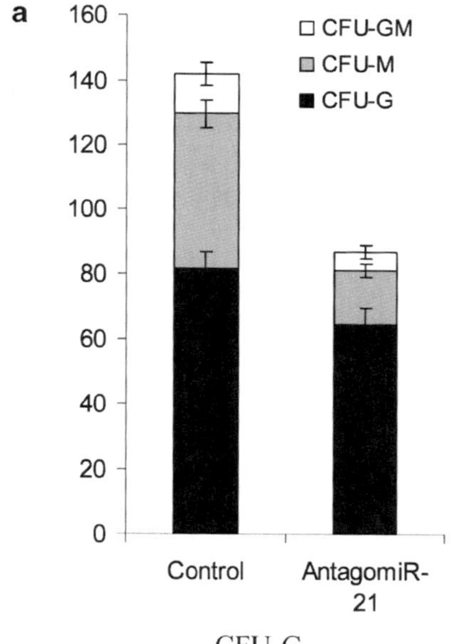

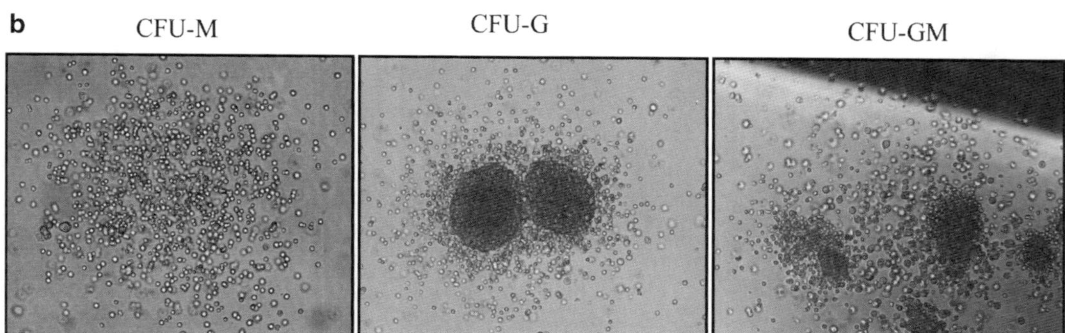

Fig. 3. Biological assay to confirm the silencing of miR-21 with antagomiRs. (**a**) Colony-forming unit from the control- and antagomiR-21-treated wild Lin-bone marrow cells showing lower total colony number, specifically reduced monocytic colonies. (**b**) Representative picture showing CFU-M, CFU-G, and CFU-GM.

5. Plate the cells on a grid plate for easy scoring and incubate in cell culture incubator.
6. Count colonies after 7–10 days, score the colonies which contain more than 50 cells, and differentiate based on their morphology (CFU-G, CFU-M, CFU-GM) (Fig. 3b). (CFU-G, progenitors of granulocytes, give rise to eosinophils, basophils, or neutrophils; CFU-M, progenitors of macrophages, give rise to macrophages; CFU-GM, progenitors, give rise to a heterogeneous population of macrophages and granulocytes.)

3.5. In Vivo Assay

For a continuous dosing, miniature infusion (osmotic) pumps provide a convenient and reliable method for controlled delivery of drugs. Osmotic pumps are extensively used to study the effects of compounds in vivo, because they eliminate frequent animal handling, repetitive injection schedules, and ensure the delivery of drugs at a constant dose.

3.5.1. Loading the antagomiRs

1. Prepare antagomiR solution of known concentration to load into the pump (ensure that solutions are at room temperature) (see Note 4).

   Example: A Model 2006 pump with a mean pumping rate of 0.15 μL/h is filled with 240 μL and implanted for 42 days. Upon removal, approximately 70–80 μL can be aspirated from the reservoir. So, one can assume that approximately 151 μL would be infused during the 42-day infusion period:

$$\text{Mean Pumping rate} \times \text{infusion duration (in hours)} = \text{Volume infused}$$

$$0.45\mu\text{L}/\text{h} \times (24\text{ h} \times 42\text{ days}) = 151\mu\text{L}$$

   This would leave a residual volume of 70–80 μL (which give an extra 20 days, 3.6 μL/day).

2. Wear gloves and carefully load antagomiRs into the pump. Use the blunt-tip needles supplied with the pumps. Do this by moving the plunger to 0.1 mL and then drawing up the fluid slowly. This ensures that all of the liquid leave the syringe (see Note 5).
3. Hold the pump in an upright position, with the exit port pointed vertically.
4. Slowly load the antagomiR to the pump by putting the loading needle to the bottom of the pump and loading until the liquid reaches the surface of the pump. Make sure that no air bubbles are in the pump.
5. Wipe off the excess solution and insert the flow moderator directly into the pump making sure that the cap is flushed with the top of the pump.

6. The insertion displaces some of the liquid from the filled pump through the exterior hole. (Moderator must be straight to work properly. This can make the pump not work.)
7. Place the pump in a 50 mL conical tube with about 20 mL of 0.9% saline (see Note 6).

3.5.2. For Subcutaneous Placement of Pump

1. Inject the mouse with 250 μL of Avertin and 30–40 μL of Brupenex and let it go to sleep.
2. Once the mouse is anesthetized, shave a patch of hair from the back on the neck right behind the ears with the electric shaver.
3. Clean the shaved area with alcohol swab.
4. Make a small insertion/cut in the skin (big enough for the pump) using the sharp scissors and use the blunt-tipped scissors to loosen the subcutaneous connective tissue apart under the skin.
5. Insert the Alzet Osmotic pump (moderator in first) into the animal and push it deep into the body to place subcutaneously.
6. Use 1–2 wound clips to close wound entirely.
7. Keep mouse warm during the recovery as long as possible.
8. Mice can be housed together with the wound clips.

Assay for Target Gene Activity: Taqman Expression

As shown in Fig. 1b, the efficacy of antagomiR-21 in silencing endogenous mature miR-21 was confirmed in the peripheral blood.

1. Collect peripheral blood from the control- and antagomiR-21-treated mice using standard bleeding technique described in an approved protocol (retro-orbital, jugular vein, tail vein).
2. Lyse the red blood cells using ACK lysis buffer and wash with cold 1× PBS.
3. To validate the miRNA expression, extract total cellular RNA from the white blood cells using TRI reagent and perform the microRNA TaqMan assay as described previously.

4. Notes

1. Mature miRNA sequences can differ between species. It is important to consider this fact for antagomiR design as well as for mature miRNA detection using specific microRNA Taqman.
2. Heat at 55°C for 15 min if necessary.

3. Thaw the MethoCult and shake the bottle vigorously to thoroughly mix the contents. Allow air bubbles to escape by placing the bottle at room temperature for at least 1 h.
4. AntagomiRs are dosed at a concentration of 80 mg/kg.
5. Make sure that you filter the antagomiR through a 0.22 μM filter to ensure the sterility. Also use sterile technique while filling, handling, and implanting Alzet Osmotic pumps.
6. Place the pumps with saline for at least 4–60 h before implantation (depends upon the pump used) in order to prime the pumps for continuous delivery of the drugs.

Acknowledgments

This work was partially supported by grants from Alex's Lemonade Stand, and NIH R21 CA142601 and R01 CA159845 (H.L.G.).

References

1. Nakasa T, Nagata Y, Yamasaki K, Ochi M (2011) A mini-review: MicroRNA in Arthritis. Physiol Genomics 43(10):566–70
2. Lewis BP, Burge CB, Bartel DP (2005) Conserved seed pairing, often flanked by adenosines, indicates that thousands of human genes are microRNA targets. Cell 120:15–20
3. Zhdanov VP (2009) Conditions of appreciable influence of microRNA on a large number of target mRNAs. Mol Biosyst 5:638–43
4. Shah AA, Meese E, Blin N (2010) Profiling of regulatory microRNA transcriptomes in various biological processes: a review. J Appl Genet 51:501–7
5. Chen CZ, Li L, Lodish HF, Bartel DP (2004) MicroRNAs modulate hematopoietic lineage differentiation. Science 303:83–6
6. Calin GA, Croce CM (2006) MicroRNA signatures in human cancers. Nat Rev Cancer 6:857–66
7. Calin GA, Sevignani C, Dumitru CD, Hyslop T, Noch E, Yendamuri S, Shimizu M, Rattan S, Bullrich F, Negrini M, Croce CM (2004) Human microRNA genes are frequently located at fragile sites and genomic regions involved in cancers. Proc Natl Acad Sci USA 101:2999–3004
8. Chen CZ, Li L, Lodish HF, Bartel DP (2004) MicroRNAs modulate hematopoietic lineage differentiation. Science 303:83–6
9. Trang P, Wiggins JF, Daige CL, Cho C, Omotola M, Brown D, Weidhaas JB, Bader AG, Slack FJ (2011) Systemic Delivery of Tumor Suppressor microRNA Mimics Using a Neutral Lipid Emulsion Inhibits Lung Tumors in Mice. Mol Ther 19(6):1116–22
10. Todesco M, Rubio-Somoza I, Paz-Ares J, Weigel D (2010) A collection of target mimics for comprehensive analysis of microRNA function in Arabidopsis thaliana. PLoS Genet 6:e1001031
11. Gentner B, Schira G, Giustacchini A, Amendola M, Brown BD, Ponzoni M, Naldini L (2009) Stable knockdown of microRNA in vivo by lentiviral vectors. Nat Methods 6:63–6
12. Scherr M, Venturini L, Eder M (2010) Lentiviral vector-mediated expression of pre-miRNAs and antagomiRs. Methods Mol Biol 614:175–85
13. Mattes J, Yang M, Foster PS (2007) Regulation of microRNA by antagomirs: a new class of pharmacological antagonists for the specific regulation of gene function? Am J Respir Cell Mol Biol 36:8–12
14. Griffiths-Jones S (2006) miRBase: the microRNA sequence database. Methods Mol Biol 342:129–38
15. Cooper SR, Taylor JK, Miraglia LJ, Dean NM (1999) Pharmacology of antisense oligonucleotide inhibitors of protein expression. Pharmacol Ther 82:427–35
16. Velu CS, Baktula AM, Grimes HL (2009) Gfi1 regulates miR-21 and miR-196b to control myelopoiesis. Blood 113:4720–8

Chapter 16

Synthesis, Conjugation, and Labeling of Multifunctional pRNA Nanoparticles for Specific Delivery of siRNA, Drugs, and Other Therapeutics to Target Cells

Peixuan Guo, Yi Shu, Daniel Binzel, and Mathieu Cinier

Abstract

RNA is unique in nanoscale fabrication due to its amazing diversity of function and structure. RNA nanoparticles can be fabricated with a level of simplicity characteristic of DNA while possessing versatile tertiary structure and catalytic function similar to that of proteins. A large variety of single stranded loops are suitable for inter- and intramolecular interactions, serving as mounting dovetails in self-assembly without the need for external linking dowels. Novel properties of RNA nanoparticles have been explored for treatment and detection of diseases and various other realms. The higher thermodynamic stability, holding of noncanonical base pairing, stronger folding due to base stacking properties, and distinctive in vivo attributes make RNA unique in comparison to DNA. Indeed, the potential application of RNA nanotechnology in therapeutics is an exciting area of research.

The use of RNAi in biomedical research has opened up new possibilities to silence or regulate the biological function of individual genes. Small interfering RNA (siRNA) has been extensively explored to genetically manipulate the expression in vitro and in vivo of particular genes identified to play a key role in cancerous or viral diseases. However, the efficient silencing of the desired gene depends upon efficient delivery of siRNA to targeted cells, as well as in vivo stability. In this chapter, we use the bacteriophage phi29 motor pRNA-derived nanocarrier as a polyvalent targeted delivery system, introduce the potential of RNA-based therapeutics using nanobiotechnology or nanotechnology methods with the fabrication and modification of pRNA nanoparticles, and highlight its potential to become a valuable research tool and viable clinical approach for gene therapy.

Key words: Bacteriophage phi29, Nanomotors, RNA nanotechnology, Nanobiotechnology, Bottom-up assembly, pRNA nanoparticle, RNAi, Cell-type specific delivery, Viral DNA packaging motor, Viral assembly

1. Introduction

Nanotechnology involves modification, engineering, and/or assembly of organized materials at the nanometer scale (1–3). RNA molecules can be designed and manipulated at a level of simplicity

Yi Zheng (ed.), *Rational Drug Design: Methods and Protocols*, Methods in Molecular Biology, vol. 928,
DOI 10.1007/978-1-62703-008-3_16, © Springer Science+Business Media New York 2012

similar to DNA (4–8) while possessing the versatility in structure, function, and enzymatic activity similar to that of proteins (9–14), making RNA an attractive candidate for nanotechnological applications (15).

1.1. RNA in Therapeutics

In recent years, more and more functional RNA molecules, naturally or artificially engineered, have been discovered. Ribozyme, antisense RNA, riboswitch, miRNA, siRNA, and aptamer have shown potential as therapeutics. Like antibodies, RNA aptamers selected by systematic evolution of ligands by exponential enrichment (SELEX) (16–20) are able to bind to specific targets, including proteins, organic compounds, and nucleic acids (21–23). The ability to recognize these specific cell surface markers through the formation of binding pockets, and the capability of internalization by the targeted cells paves a new way for targeted delivery (24–29). Secondly, in 1998, Fire and Mello discovered RNA interference (RNAi), which can regulate gene expression on a posttranscriptional level (30). These RNA molecules have significant therapeutic potential and are capable of regulating gene function by intercepting and cleaving mRNA or the genome of RNA viruses. The discovery of RNAi has heightened interest in RNA therapeutics (30), since several RNA-based therapeutic approaches using small interfering RNA (siRNA)(24, 31–42), ribozyme, and antisense RNA (9–14, 43, 44) have been shown to downregulate specific gene expression in cancerous and viral-infected cells. The successful application of RNAi-based therapeutics for the treatment of cancer and infectious diseases requires several features: (1) specific delivery to cells, (2) capability of entering cells, (3) surviving degradation by nucleases within the cell, (4) trafficking into the appropriate cell compartments, (5) correct folding of siRNA or ribozyme in the cell if fused to a carrier, (6) the release and incorporation of siRNA into RNA-Induced Silencing Complex (RISC), and (7) low toxicity and low immunogenicity. RNA nanoparticles derived from phi29 pRNA have been demonstrated to fulfill most of these features (45–48).

1.2. pRNA Nanoparticle as a Multivalent Nanocarrier

To meet the specific requirements related to the treatment of cancer via RNA-based therapeutics, a RNA-based siRNA vector was inspired by the molecular machinery within the phi29 bacteriophage DNA packaging motor (Fig. 1d) (4, 49). The motor contains six RNA molecules, 117-nucleotide in length, which form a hexameric ring by "hand-in-hand" interactions (Fig. 1b, d) (4, 50). Extensive structural and functional studies of the motor revealed that the RNA hexamer plays an essential role in genomic DNA translocation and packaging (Fig. 1d). Thus, it has since been called "packaging" RNA (pRNA).

1.2.1. pRNA Structure and Self-Assembling Property

Bacterial virus phi29 DNA packaging RNA (pRNA) molecules contain an intermolecular interaction central domain and a helical

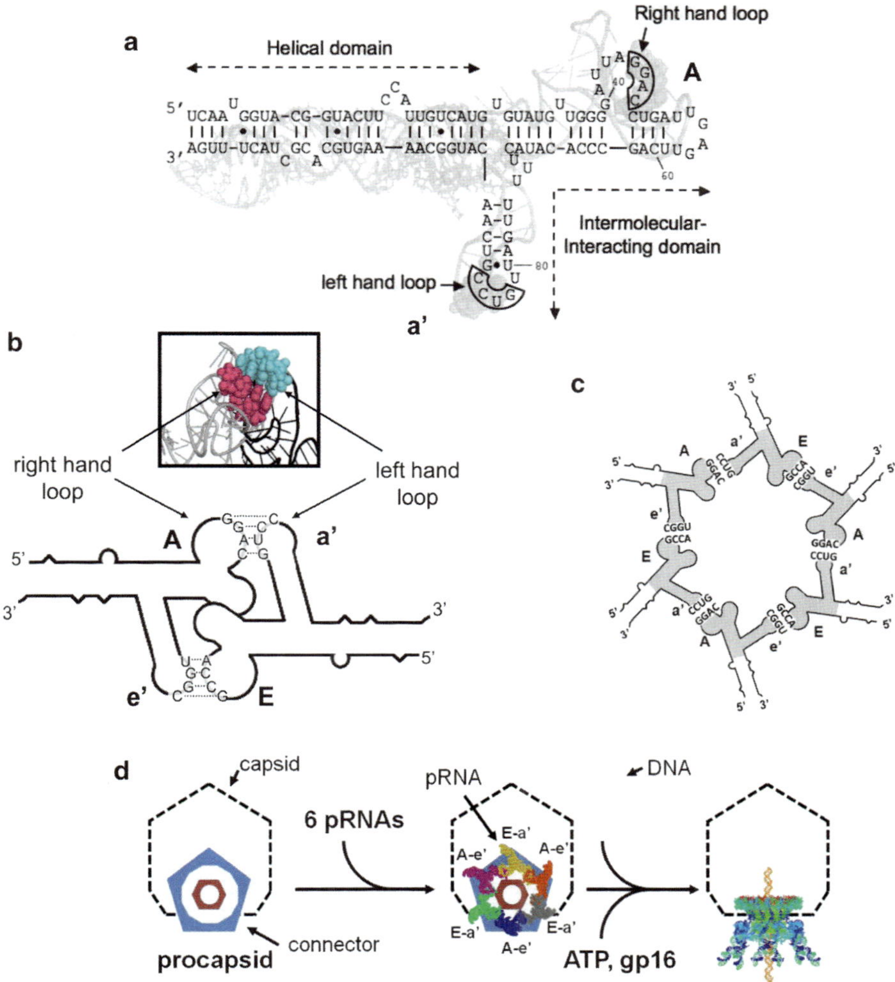

Fig. 1. Sequence and secondary structure of phi29 DNA packaging RNA (pRNA). (**a**) Superposition of the 2D and 3D structure of the pRNA Aa′. The four bases in the *right* and *left* loops, which are responsible for inter-RNA interactions are *boxed*. (**b**) Depiction of pRNA dinner formation and 3D model of the hand-in-hand interaction. (**c**) pRNA hexamer. (**d**) Packaging of phi29 DNA thorough the motor geared by six pRNA (from ref. 61, with permission).

domain as the 5′/3′ paired region (Fig. 1a). Via the interaction of two interlocking left and right loops, the pRNA molecules form dimers, trimers, hexamers, and other patterned superstructures (6). This property of forming self-assembled nanostructure makes pRNA ideal building blocks for bottom-up assembly. The ability to form pRNA multimers is not affected by 5′ or 3′-end truncation before residue #23 and after residue #97 (51). Thus, end conjugation of pRNA with chemical moiety fusing pRNA with a receptor-binding RNA aptamer, siRNA, or ribozyme would not disturb pRNA dimer formation or interfere with the function of inserted moieties (13, 14, 24, 42, 52).

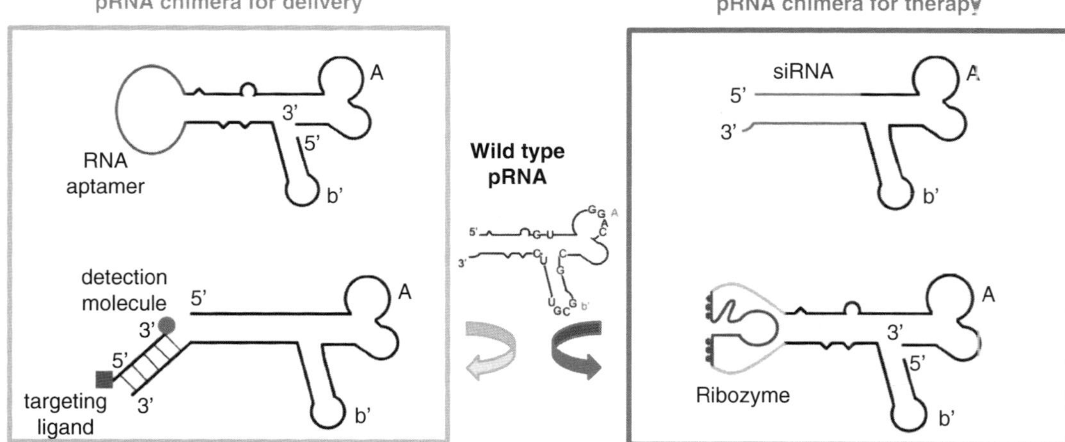

Fig. 2. The construction of pRNA chimeras including therapeutic position like pRNA/siRNA and pRNA/ribozyme as well as delivery portion such as pRNA/aptamer and pRNA conjugated with chemical ligands (from ref. 46, with permission).

1.2.2. Design of Multivalent pRNA Nanoparticle for Targeted Drug Delivery

The self-assembly of pRNA nanoparticles is a prominent bottom-up approach for nanoparticles fabrication and represents an important idea that biomolecules can successfully integrate into nanotechnology (6, 24, 52–54). Recently, the phi29 pRNA monomer has been designed to carry a variety of therapeutic or reporter agents including siRNA (24, 46, 52), hammerhead ribozymes, receptor-binding RNA aptamers, targeting ligands such as folate, as well as other small molecules such as fluorophore and biotin (Fig. 2). Phi29 pRNA monomers with different cargo molecules can be subsequently used in the assembly of dimers, trimers, or hexamers, which enables the use of phi29 derived nanocarriers as polyvalent delivery vehicles for the development of drugs and treatment of various diseases (Fig. 3). Dimeric, trimeric, and potentially hexameric pRNA nanoparticles can be easily formed via the interactions of rationally designed pRNA monomers with interlocking left and right loops (Table 1). By using a bottom-up approach, we fabricated multivalent pRNA nanoparticles with a precise control of the stoichiometry of the different functional elements, such as a dimer comprised of one pRNA building block harboring siRNA and another pRNA building block carrying a receptor-binding ligand or aptamer. We achieved the delivery of siRNA via the receptor-binding ligand to the desired cells as reported by the fluorescent marker, thus enabling targeted delivery. Incubation with these targeted pRNA/siRNA nanoparticles resulted in the binding and co-entry of the divalent or trivalent therapeutic particles into cells, subsequently modulating the apoptosis of cancer cells (24, 42, 52, 55, 56).

1.2.3. Advantage of pRNA Nanoparticle

It is commonly accepted that RNAs can induce a lower detectable antibody response compared to proteins if the RNA does not form

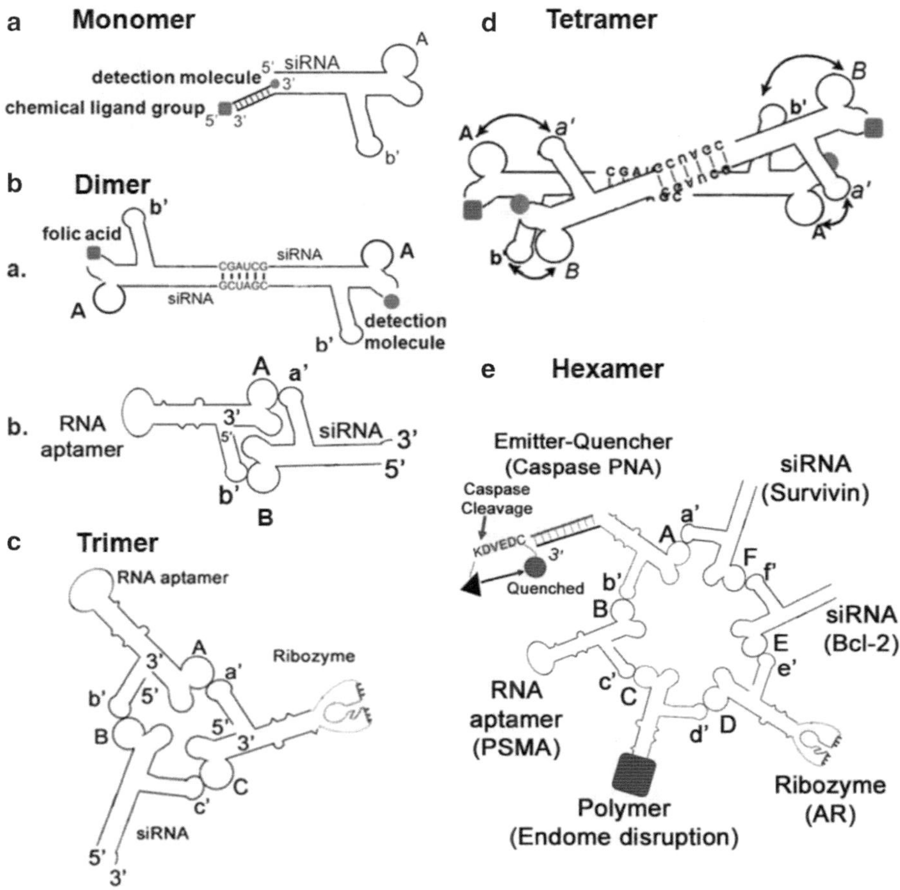

Fig. 3. Illustration of constructed therapeutic pRNA multimers. (**a**) pRNA chimeric monomer with 3′-end annealed oligo which conjugates the detection molecule and chemical ligand at the same time. (**b**) (a) Twin dimer formed through palindrome sequence; (b) Dimer formation through loop–loop interaction. (**c**) Trimeric RNA chimera, (**d**) Tetrameric pRNA chimera, and (**e**) Hexameric pRNA particle (from ref. 46, with permission).

complex with proteins. The use of RNA as a delivery vehicle has the potential to reduce the antibody response, thus allowing repeated long-term drug administration for chronic diseases. In addition, pRNA-derived nanoparticles have a small size (20–40 nm) and narrow size distribution which makes them particularly suited for in vivo systemic delivery. The ability to easily functionalize pRNA multimers for target-specific localization and the feasibility of multiple therapeutic agent delivery using a single vehicle are other significant advantages of the pRNA approach. Each hexameric pRNA nanoparticle offers six positions available to carry molecules for cell recognition, therapy, and/or detection. Moreover, the resistance of the pRNA nanoparticle against RNase degradation can be significantly increased with the incorporation of 2′-F modifications in the pRNA backbone (45). A variety of other molecules such as heavy metals, quantum dots, fluorescent beads, or radioisotopes

Table 1
Loop sequences of pRNA that can be used for self-assembly of pRNA nanoparticle via hand-in-hand interaction

A	5′-GGAC-3′	F	5′-AGAC-3′
	||||		||||
a′	3′-CCUG-5′	f	3′-UCUG-5′
B	5′-ACGC-3′	I	5′-AACC-3′
	||||		||||
b′	3′-UGCG-5′	i′	3′-UUGG-5′
C	5′-GACA-3′	J	5′-CCGU-3′
	||||		||||
c′	3′-CUGU-5′	j′	3′-GGCA-5′
D	5′-AGGC-3′	K	5′-UCCU-3′
	||||		||||
d′	3′-UCCG-5′	k′	3′-AGGA-5′
E	5′-GCCA-3′	P	5′-UCCU-3′
	||||		||||
e′	3′-CGGU-5′	p′	3′-AGGA-5′

can also be conjugated for detection via imaging of cancer signatures at different developmental stages. Therapeutic efficiency can be improved by conjugating pRNA to endosmolytic chemicals, which can trigger the release of the internalized therapeutic reagents from the endosome (for review see refs. (3, 15, 46, 57)).

1.3. Functionality of pRNA-Based Nanoparticles

Different strategies have been applied to the construction of functional pRNA units that can be further assembled into one multivalent nanoparticle through the bottom-up approach (46). pRNA nanoparticle construction involves (1) chemical modification of nucleotides, (2) 5′-end labeling of subunits and circular permutation, (3) construction of chimeric pRNA, and (4) cross-linking of modules.

1.3.1. Chemical Modification of Nucleotides to Make Stable RNA Nanoparticles

Chemical instability of RNA is one of the major challenges in RNA therapeutics. Standing in awe of the sensitivity of RNA to RNase degradation has made many scientists flinch away from the studies in RNA nanotechnology. The 2′-F (2′-fluoro) RNA retained its property for correct folding in dimer formation, appropriate structure in procapsid binding, and biological activity in gearing the phi29 nanomotor to package viral DNA and producing infectious

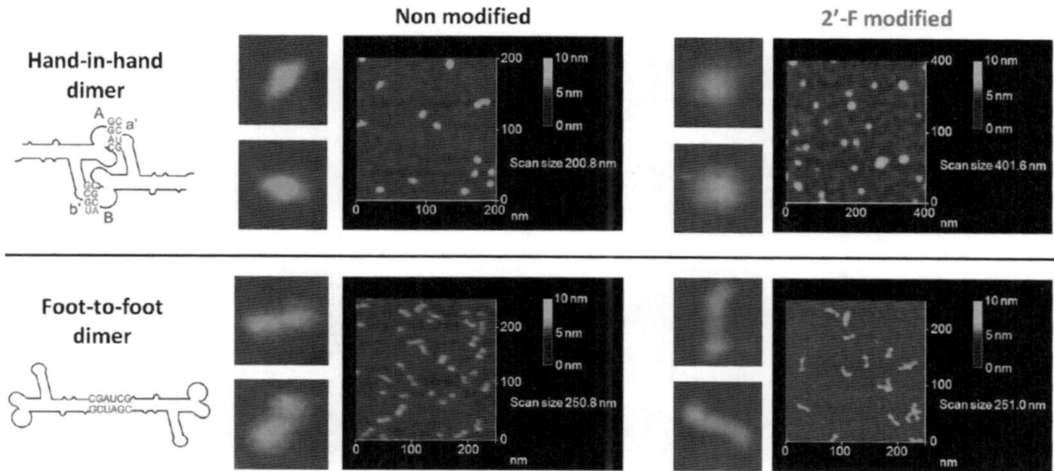

Fig. 4. Atomic force microscopy images showing hand-in-hand and foot-to-foot dimer nanoparticle of nonmodified pRNA and 2′-F-C/U pRNA, respectively (from ref. 61, with permission).

viral particles. Our results demonstrate that it is practical to produce RNase-resistant, biologically active, and stable RNA for application in nanotechnology. Many chemically modified nucleosides are now commercially available such as 2′-modified ribonucleoside where the 2′-OH of the ribose have been methylated (58) or substituted with either a fluorine (45, 59) or an amine (60, 61). Different T7 RNA polymerase mutants were constructed and shown to recognize and permit the incorporation of those 2′-modified triphosphate ribonucleotide such as 2′-OMe, 2′-F, 2′-NH_2, 2′-N_3 into the RNA chain (45, 62–65). This strategy has been reported many times to increase the RNase resistance without sacrificing the in vivo potency of siRNA molecules (66–69).

Recently, we have reported that the substitution of all pyrimidine in the pRNA sequence with their 2′-F derivatives did not affect the folding and shape of the pRNA molecule as shown by AFM images (Fig. 4). Also, 2′-F modified pRNA molecules maintain their ability to assemble nanoparticles via hand-in-hand interaction and the respective nanoparticles have shown improved resistance to RNase degradation compared to unmodified RNA (45).

1.3.2. 5′-End Labeling of pRNA Subunits by Circular Permutation

Although it is not difficult to incorporate a single label during solid phase synthesis of short RNA, synthesis of long RNA relies on the enzymatic methods, with the single labeling of the RNA hard to achieve while maintaining high labeling efficiency. To overcome this challenge and label the longer RNA with a single functional group, AMP and GMP derivatives, which have been demonstrated to be efficient initiators in RNA transcription by T7 RNA polymerase but cannot be used in the elongation step, were designed (42, 70–72). Therefore, efficient labeling of RNA molecules at the specific 5′ position can be readily achieved by either one-step transcription

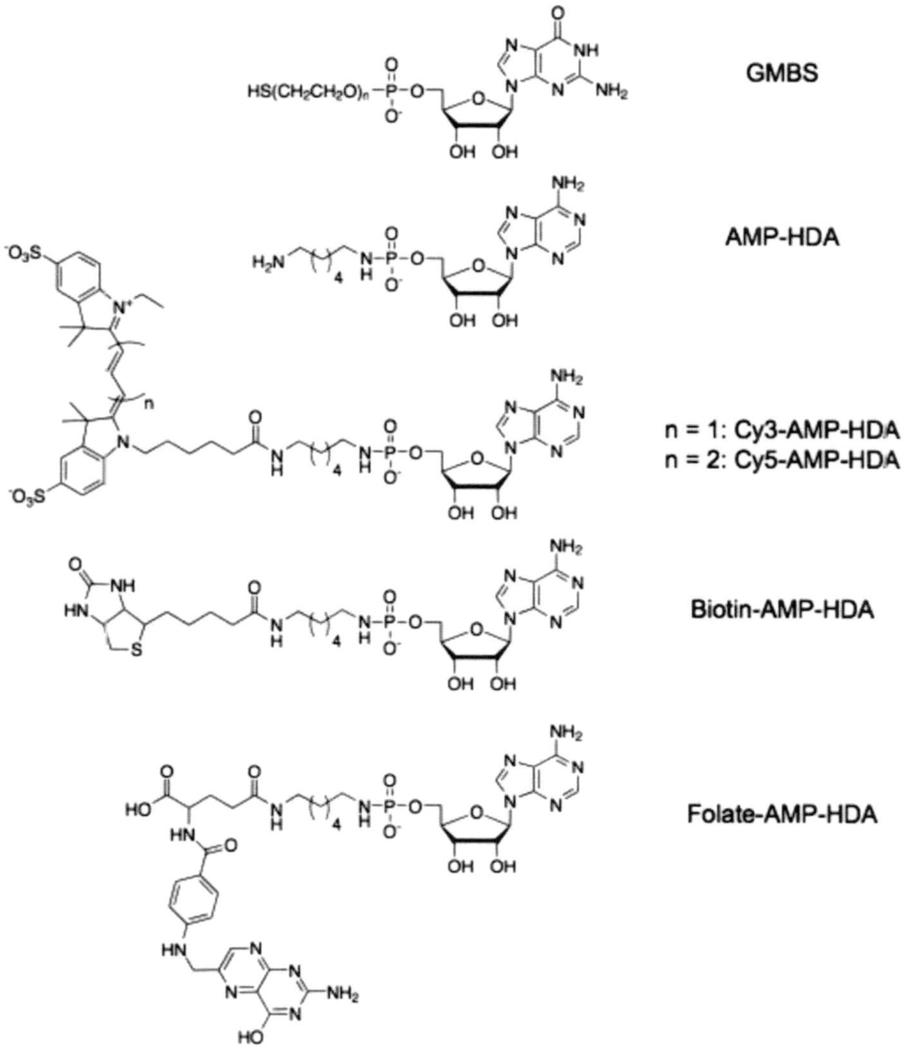

Fig. 5. The chemical structure of the synthesized AMP and GMP derivatives (from ref. 46, with permission).

initiation or a two-step procedure of transcription and posttranscriptional modification. Then, labeling in the desired position can be achieved by using circular permutation strategy (73–75).

5′-End Labeling with AMP and GMP Derivatives

Variations of AMP and GMP derivatives were synthesized by conjugating several chemical groups to AMP or GMP through linker molecules with established chemistry (42, 70–72, 76–79) (Fig. 5).

The AMP-hexanediamine (HDA) (70, 72) can be obtained easily from the direct coupling of AMP with HDA in presence of 1-Ethyl-3-[3-dimethylaminopropyl]carbodiimide hydrochloride (EDC) (72) followed by a subsequent purification by reverse phase HPLC. The free amine-reactive moiety can be then used for further functionality by common coupling reactions with *N*-hydroxysuccinimide (NHS) activated compounds. Fluorescent dyes (71, 72) (FITC, Cy5, Cy3), targeting moieties (Folic acid (42)), and biotin (72), were successfully coupled to AMP-HDA and the resulting derivatives were shown to be efficient for the construction of single labeled pRNA molecules (80). Alternatively, these NHS activated compounds can be coupled posttranscriptionally to the end NH_2 group of single labeled RNA fragments with AMP-HDA.

Construction of Active Circularly Permuted pRNA (cp-pRNA) (74, 75)

Utilizing 5′ to 3′-end ligation, Pan et al. were able to construct circular tRNA, which was subsequently cleaved with limited alkaline hydrolysis to generate one random break per molecule (81) and generated RNA molecules that were used for structural analysis. Similarly, Nolan et al. synthesized tRNAs with native 5′ and 3′ ends linked by a synthetic loop and new termini in the interior of the native sequence (82). The feasibility of constructing circularly permuted RNAs rests on the close proximity of the native RNA 5′ and 3′ ends. The 5′- and 3′-ends of the phi29 pRNA being arranged closely (Fig. 1a), it was possible to construct a series of pRNA molecules with circular permutations (74, 75).

To construct cp-pRNA, two tandem pRNA-coding sequences separated by a 3-base or 17-base loop sequence have been cloned into a plasmid (Fig. 6). Polymerase chain reaction (PCR) primer pairs, complementary to various locations within pRNA-coding sequences, could be designed to synthesize PCR fragments for

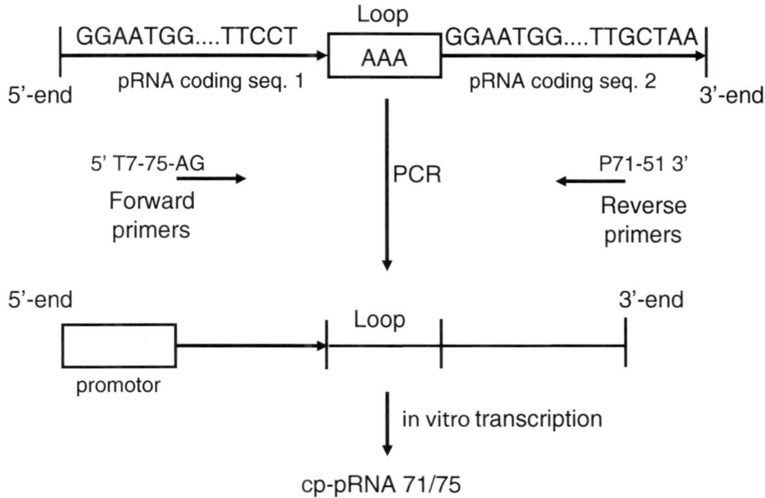

Fig. 6. Construction of tandem DNA templates for the synthesis of circularly permuted pRNA (cp-pRNA). Two pRNA-coding sequences are cloned into a plasmid with head-to-tail linkage via three nucleotides AAA. PCR primers, such as 5′T7-75-AG/P71-51 3′ containing T7 promoter, are used to generate a DNA fragment for RNA transcription. *Arrows* indicate positions of new 5′/3′ termini that have been used (adapted from ref. 72, with permission).

transcription of cp-pRNAs. It has been shown that neither a small nor a large linker loop interferes with the biological activity of the pRNA molecule (74). Additionally, most of the bases, especially the nonessential bases, can be used as new termini for constructing active cp-pRNA (74, 75). The nascent 5′/3′ end of the pRNA can be relocated to nt 71 and 75, which is protected the 5′/3′ end from exonuclease digestion.

1.3.3. Construction of Chimeric pRNA

The pRNA double-stranded 5′/3′ end helical domain and intermolecular interacting domain fold independently of each other. Complementary modification studies have revealed that altering the primary sequences of any nucleotide of the helical region does not affect pRNA structure and folding as long as the two strands are paired (75). Thus, end conjugation of pRNA with chemical moiety of fusing pRNA with a receptor-binding RNA aptamer, siRNA, or ribozyme appears to be feasible; and will not interfere with the function of inserted moieties (13, 14, 24, 42, 52).

pRNA Chimera Harboring siRNA (46) (Figs. 2 and 3)

Extensive studies revealed that siRNA is a 21–23 nt RNA duplex (33, 34, 37, 83). Thus, it was possible to replace the helical region in pRNA with double-stranded siRNA. A variety of chimeric pRNAs with different targets were constructed to carry siRNA connected to nt# 29 and 91 of the pRNA (Fig. 1a), which successfully inhibited target gene expression (42, 52, 56). A model involving the anti-apoptotic survivin gene as a proof of concept, where the specific knockdown of the survivin gene can drive cancer cells to undergo apoptosis, reduce cell viability, and prevent tumorigenesis in xenograft mice models (24, 52).

pRNA Chimera Harboring Ribozyme (46) (Figs. 2 and 3)

Using the circular permutation approach pRNA (74, 75), almost any nucleotide of the entire pRNA can serve as either the new 5′ or 3′-end of the RNA monomer. Connecting the pRNA 5′/3′ ends with variable sequences did not disturb its folding and function (13, 14, 24, 42, 52). These unique features, which help prevent exonuclease degradation and misfolding, make pRNA an ideal vector to carry therapeutic RNA such as ribozymes. A pRNA-based vector was designed to carry hammerhead ribozymes (13). The pRNA/ribozyme(survivin), which targeted the anti-apoptosis factor survivin to downregulate genes involved in tumor development and progression, was also shown to suppress survivin expression and initiate apoptosis (14).

pRNA Chimera Harboring Aptamer (46) (Figs. 2 and 3)

In vitro SELEX (22, 23) of RNA aptamers has become a powerful tool for selecting RNA molecules that target specific cell surface receptor. Such aptamers became a possible choice for specific cell targeting and utilized in the pRNA design. Aptamers were linked to the 3′ and 5′ end of cp-pRNA with 5′/3′ end relocated to nt 71 and 75. To facilitate independent folding, poly U or poly A linkers were placed between the pRNA and the aptamer.

1.3.4. Cross-linking

pRNA molecules self-assemble in nanoparticles of predefined shape. However, dissociation of the loop can occur at low concentration or when the concentration of Mg^{2+} is reduced. To stabilize the pRNA nanoparticle, cross-linking agents such as Azido (79) or Psoralen derivative (84) have been used. Psoralen can intercalate into RNA or DNA helices and, upon irradiation with 320–400 nm light, freeze the uridine of RNA or the thymidine of DNA that are in close proximity (helix or pseudoknot) by covalent attachment (85, 86). When necessary, the sites of cross-links can be determined by primer extension (84) and/or Mung bean nuclease treatment (87). We recommend the psoralen derivative, AMT (4′-aminomethyl-4,5′,8-trimethyl psoralen), due to its solubility (84). One advantage of psoralen cross-linking is that it exclusively cross-links RNA and DNA, but not protein.

1.4. Concluding Remarks

The use of small RNA in gene therapy was significantly hampered due to the difficulties of producing a safe and efficient system to recognize and target specific cells for delivery. The strength of using phi29 pRNA as a delivery vehicle relies on its ability to easily form stable multimers, which could be manipulated, engineered via sequence-controlled and self-assembly (4, 88, 89). This particular system provides unprecedented versatility in constructing polyvalent delivery vehicles by separately constructing individual pRNA subunits with various cargos and mixing them together in any desired combination (3).

These nanoparticles can carry multiple components, including molecules for specific cell recognition, image detection, endosome disruption, and therapeutic treatment (Fig. 6). One subunit of the deliverable RNA nanoparticle (dimer, trimer, or tetramer) can be modified to carry a RNA aptamer or other ligands that binds to a specific cell surface receptor, thereby inducing receptor-mediated endocytosis. The second subunit can carry reporting molecules for the evaluation of cell binding and entry. The third subunit can be reengineered to carry components that can be used to enhance endosome disruption so that the therapeutic molecules are released. The fourth (or fifth and sixth, if needed) subunit of the RNA nanoparticle can carry therapeutic siRNA, ribozyme RNA, antisense RNA, miRNA, or other drugs to be delivered.

The assembled nanoscale particles harboring functional moieties offer many advantages over the numerous other anticancer delivery platforms under development such as polyvalent delivery, controllable structure and precise stoichiometry, nanoscale size, targeted delivery, long half-life in vivo, no induction of interferon or toll-like immunity, low or no toxicity, and noninduction of an antibody response to ensure repeated treatments. Consequently, pRNA nanoparticle delivery will potentially lead to an innovative therapeutic strategy for developing an effective and safe treatment for cancer and other diseases.

2. Materials (Note 1)

2.1. Synthesis of pRNA Monomer

1. 6× loading buffer: 40% Sucrose, 0.1% (w/v) xylene cyanol FF, and 0.1% (w/v) bromophenol blue.
2. 2× TBE loading buffer: 95% formamide, 18 mM EDTA, 0.025% SDS, 0.025% bromophenol blue, and 0.025% xylene xyanol.
3. Primers (100 μM each): Regular pRNA (Forward: 5′-**TAATACGACTCACTATATAG**AATGGTACGGTACTTCC; Reverse: 5′-GGAAAGTAGCGTGCACTTTTGCCATGATTGA), cp-pRNA 71/75 (Forward: 5′- **TAATACGACTCACTATTAG**TTGATTGTCCGTC; Reverse: 5′-GTATGTGGGCTGAACTCAATC), pRNA with flanked sequence (Forward: 5′-**TAATACGACTCACTATAG**GTCATGTGTATGTTGGG; Reverse: 5′-CTCCCGGCCGCCATGGCCGCGGGATGGCCATGATTGAC). The sequence in bold contains the T7 promoter.
4. 0.5 M EDTA, pH 8: Dissolve 186.1 g of EDTA in 800 ml of DI water. Stir vigorously and adjust pH to 8 with NaOH (~20 g). Add DI water to a total volume of 1,000 ml.
5. TAE buffer: 40 mM Tris–Acetate, 1 mM EDTA. Dissolve 4.84 g of tris-base in 200 ml of DI water. Add 1.14 ml of glacial acetic acid and 2 ml of 0.5 M EDTA solution pH 8. Add DI water to a total of 1,000 ml.
6. 0.05% (v/v) DEPC aqueous solution: Mix 0.05 ml of DEPC in 99.5 ml of DI water and shake vigorously the solution. Incubate at 37°C overnight and then autoclave the solution to remove DEPC.
7. 3 M NaOAc, pH 5.2: Dissolve 246 g of sodium acetate in 800 ml of DI water. Adjust the solution pH to 5.2 with acetic acid. Add DI water to a total volume of 1,000 ml.
8. 2% Agarose–Synergel: Dissolve 0.64 g of Synergel in 2 ml of ethanol. Add 100 ml of TAE buffer and 0.7 g of agarose. Heat the suspension until agarose is dissolved completely, add 1 μl of EtBr for 25 ml gel, then pour before the solution cools down.
9. 5× transcription buffer: 400 mM HEPES–KOH pH 7.5, 120 mM $MgCl_2$, 10 mM spermidine, and 200 mM DTT. Dissolve 0.95 g of HEPES, 0.31 g of DTT, 0.025 g of spermidine, and 0.24 g of $Mg\ Cl_2$ in 10 ml of DI water. Add KOH to the solution to adjust the pH at 7.5.
10. NTPs mix, 25 mM each: Mix equal amount of 100 mM ATP, GTP, CTP, and UTP.
11. 100 mM DTT solution: Dissolve 1.54 g of DTT in 100 ml of DI water.

12. 5× TBE buffer: Dissolve 54 g of tris-base in 800 ml of DI water. Add 27.5 g of boric acid and 20 ml of 0.5 M EDTA solution. Add DI water to a total volume of 1,000 ml.
13. 1× TBE buffer: 89 mM Tris–borate pH 8, 2 mM EDTA. Dilute 200 ml of 5× TBE buffer with 800 of DI water.
14. 10% APS (ammonium persulfate): Dissolve 10 g of APS in 100 ml of DI water.
15. 8% Polyacrylamide gel with 8 M of Urea: Place a clean 1,000 ml glass beaker containing a large magnetic stir bar (1.3 cm × 7.6 cm) over a magnetic stirrer. Add 480 g of urea, 200 ml of 40% (v/v) acrylamide/bisacrylamide solution (29:1), and 200 ml of 5× TBE buffer. Add DI water to a total volume of 1,000 ml. Stir continuously stir until solid urea dissolves completely. Filter the solution on a glass wool plug. Add 6 μl of TEMED and 50 μl of 10% APS solution freshly to 5 ml of 8% polyacrylamide/8 M urea solution to prepare the gel.
16. Elution buffer: 0.5 mM NH_4OAc, 0.1 mM EDTA, 0.1% SDS, and 0.5 mM $MgCl_2$. Dissolve 3.85 g of NH_4OAc, 2.92 g of EDTA, 0.1 g of SDS, and 4.76 g of $MgCl_2$ in 100 ml of 0.05% DEPC water, autoclave and store at room temperature.
17. 2′-F transcription buffer 10×: 400 mM Tris–acetate pH 8.0, 50 mM DTT, 10 mM EDTA, 100 mM Mg-acetate, 5 mM $MnCl_2$, and 80 mM spermidine. Dissolve 4.85 g of tris-base into 10 ml of DI water and adjust pH to 8.0 with acetic acid to make 4 M Tris–acetate buffer. Then, dissolve 0.077 g of DTT, 0.037 g of EDTA·Na·$2H_2O$, 0.215 g of $MgOAc_2·4H_2O$, 0.08 g of $MnCl_2·4H_2O$, and 0.204 g of spermidine in 9 ml of DI water and add 1 ml of 4 M Tris–acetate buffer. Filter with 0.22 μm filter and store at −20°C.

2.2. 5′-End Single Molecule Labeling of pRNA Through One-Step Transcription Using AMP-HDA Derivatives

1. 6 N HCl solution: Pour 50 ml of HCl concentrate in 50 ml of DI water.
2. 2 N HCl solution: Dilute 20 ml of 6 N HCl solution with 40 ml of DI water.
3. 1 M $NaHCO_3$, pH 8: Dissolve 4.2 g of $NaHCO_3$ in 30 ml of DI water. Adjust pH to 8 with 6 N HCl. Add DI water to 50 ml.

2.3. Construction of pRNA/Folate Conjugate: Synthesis of DNA-Folate Conjugate and Annealing with pRNA

1. NH_2-DNA oligonucleotide (IDT): 5′/5AmMC6/CTCCCG GCCGCCATGGCCGCGGGATT.
2. 4× TB: 89 mM tris-base, 200 mM borate. Dissolve 43 g of tris-base in 200 ml of DI water. Add 50 g of boric acid. Add DI water to a total volume of 1,000 ml.
3. 8% PAGE gel in TBE: Dilute 1,250 μl of 4× TB, 50 μl of 0.5 M EDTA (Subheading 2.1), and 1,000 μl of 40% (v/v) acrylamide/bisacrylamide solution (29:1) in 2,700 μl of DI

water. Add 6 μl of TEMED and 50 μl of 10% APS solution (Subheading 2.1) freshly to prepare the gel.

4. 10× annealing buffer: 500 mM Tris–HCl pH 7.5, 500 mM NaCl, and 10 mM EDTA. Dissolve 6.05 g of tris-base, 2.92 g of NaCl, and 0.37 g of EDTA in 60 ml of DI water. Adjust pH to 7.5 with 6 N HCl. Add DI water to a total volume of 100 ml.
5. Buffer A: Tris 0.1 M pH 7, EDTA 0.2 mM, and NaOAc 0.4 M. Dissolve 12.11 g of tris-base, 55.23 g of NaOAc, and 0.074 g of EDTA in 800 ml of DI water. Adjust pH to 7.5 with 6 N HCl. Add DI water to a total volume of 1,000 ml.
6. Buffer B: Tris 0.1 M pH 7, EDTA 0.2 mM, NaOAc 1.5 M. Dissolve 12.11 g of tris-base, 204.12 g of NaOAc, and 0.074 g of EDTA in 600 ml of DI water. Adjust pH to 7.5 with 6 N HCl. Add DI water to a total volume of 1,000 ml.

2.4. Construction of pRNA Nanoparticle

1. 2 M $MgCl_2$: Dissolve 406.6 g of $MgCl_2$ in a total volume of 1,000 ml of DI water.
2. 0.5 M $MgCl_2$: Dilute 250 ml of 2 M $MgCl_2$ with 750 ml of DI water.
3. TBM: Dilute 250 ml of 4× TB and 2.5 ml of 2 M $MgCl_2$ in a total volume of 1,000 ml of DI water.
4. 8% PAGE gel in TBM: Dilute 1,250 μl of 4× TB, 50 μl of 0.5 M $MgCl_2$, and 1,000 μl of 40% (v/v) acrylamide/bisacrylamide solution (29:1) in 2,700 μl of DI water. Add 6 μl of TEMED and 50 μl of 10% APS solution freshly to prepare the gel.

2.5. Cross-linking

1. Stock AMT solution (0.2 mg/ml): Dissolve 2 mg of AMT in 10 ml of DI water. Store in –20°C and avoid lights.
2. (10×)AMT solution (200 ng/ml): Dilute 1 μl of AMT solution (0.2 mg/ml) in 999 μl of DI water.

3. Methods (Note 3)

3.1. Synthesis of pRNA Monomer

Methods for the construction of the DNA template of the pRNA mutants have been described previously (75) and will be not explicated here. From these constructions, the same protocol can be used for the synthesis of various pRNA mutants. To simplify the description of RNA construction and multimer assembly, uppercase letters will be used to represent the right hand loop of pRNA and lowercase letters to represent the left hand loop (Fig. 1). The same letters in upper and lower cases indicate complementary sequences for loop–loop interaction, while different letters indicate noncomplementary loops. For example, pRNA A-b′ represents pRNA with a noncomplementary right loop A ($^{5'}G_{45}G_{46}A_{47}C_{48}$)

and a left loop b′ ($^{3'}U_{85}G_{84}C_{83}G_{82}$). Dimer formation with pRNA Ab′ requires a right loop B ($^{5'}A_{45}C_{46}G_{47}C_{48}$) and left loop a′ ($^{5'}C_{45}C_{46}U_{47}G_{48}$) of pRNA Ba′ (38). Sequences of the different loops used are given in Table 1.

3.1.1. Amplification of the DNA Template by PCR

1. In a final volume of 100 μl of reaction system, mix 56 μl of 0.05% DEPC water, 20 μl of 5× GoTaq Flexi Buffer, 10 μl of 25 mM $MgCl_2$ solution, 8 μl of 2.5 mM dNTPs solution, and 2 μl of each primer, and then add 1 μl of DNA template and 1 μl of GoTaq polymerase.
2. Use the thermocycler as follow: 95°C 5 min, 22× (94°C 1 min, 55°C 2 min, 72°C 30 s), 72°C 5 min.
3. Precipitate overnight at −20°C the PCR product after addition of 2.5 volume of 100% ethanol and 1/10 volume of 3 M NaOAc.
4. Centrifuge (30 min, 16,500 ×*g*), remove the supernatant, wash the pellet with 70% ethanol and then speed vacuum dry for 5 min.
5. Redissolve the pellet in 15 μl of 0.05% DEPC water, add 3 μl of 6× loading buffer and run in a 2% agarose–Synergel gel in 1× TAE buffer (120 V).
6. Cut the band of interest under UV light, and extract the PCR DNA template from the gel using the QIAEX II gel extraction kit. Concentrate if necessary to reach at least 0.5 μg/μl of amplified DNA template.

3.1.2. In Vitro Transcription of the pRNA

1. In a final volume of 50 μl reaction system, mix 11 μl of 0.05% DEPC water, 2 μg of DNA template, 10 μl of 5× transcription buffer, 10 μl of 25 mM NTPs solution, and 5 μl of 100 mM DTT, and then add 10 μl of T7 RNA polymerase.
2. Incubate for 4 h in a water bath at 37°C.
3. End the transcription process by adding 2 μl of RNase free DNase I and incubating at 37°C for 15 min.
4. Dissolve in 15 μl TBE, add 15 μl of 2× loading buffer and load the sample in 8% polyacrylamide gel with 8 M urea in TBE (4°C, 100 V).
5. Cut the band of interest under UV light and elute the pRNA from the gel over 1.5 h at 37°C in the elution buffer.
6. Precipitate overnight at −20°C the pRNA after addition of 2.5 volume of 100% ethanol and 1/10 volume of 3 M NaOAc.
7. Centrifuge (30 min, 16,500 ×*g*), remove the supernatant, and wash the pellet with 70% Ethanol (repeat twice).
8. Dry the pellet using speed vacuum for 20 min and store the pRNA in −20°C for short use or −80°C for longer time.

3.1.3. In Vitro Transcription of 2′-F Modified pRNA (Note 2)

1. The purified PCR DNA fragments (cf. Subheading 3.1.1) are used as templates to synthesize the 2′-F modified pRNA by in vitro transcription using Y639F mutant T7 RNA polymerase.
2. In a final volume of 50 μl reaction system, mix 5 μl of 0.05% DEPC water, 2 μg of DNA template, 5 μl of 10× 2′-F transcription buffer, 5 μl of 100 mM DTT, 5 μl of 2′-F CTP, 5 μl of 2′-F UTP, 5 μl of ATP, and 5 μl of GTP, and then add 5 μl of Y639F mutant T7 RNA polymerase.
3. Incubate for overnight in a water bath at 37°C.
4. Repeat steps 3–8 of Subheading 3.1.2.

3.1.4. Circular Permutated pRNA (cp-pRNA)

To construct cp-pRNA, two tandem pRNA-coding sequences separated by a 3-base or 17-base loop sequence have been cloned into a plasmid (Fig. 6). PCR primer pairs, such as 5′T7-75-AG(a′)/3′P71-51, complementary to various locations within pRNA-coding sequences, could be designed to synthesize PCR fragments for transcription of cp-pRNAs. PCR DNA template preparation and RNA transcription are proceed with the same protocol as described in Subheadings 3.1.1 and 3.1.2.

3.2. 5′-End Single Molecule Labeling of pRNA Through One-step Transcription Using AMP-HDA Derivatives (Note 4)

Methods for 5′-end labeling of pRNA with a single chemical group have been previously reported (79, 80). We describe here the synthesis of AMP-HDA derivative with the example of the synthesis of AMP-HDA-Cy5. Similar protocol can be used for the conjugation of other NHS activated molecule to AMP-HDA in order to lead to 5′-end labeling of RNA molecule by in vitro transcription.

3.2.1. Synthesis of AMP-HDA and Conjugation with Cy5 NHS Ester

1. In a 25 ml round bottom flask, add AMP (186 mg, 0.5 mmol) and HDA (464 mg, 4 mmol) to 10 ml of deionized water.
2. Adjust the pH to be 6.5 with 2 N HCl solution.
3. During agitation, add EDC (480 mg, 2.5 mmol) stepwise with interval of 2 min. Maintain the pH around 6.5.
4. Let agitate two more hours.
5. Purify through reverse phase HPLC (Reverse phase C8 column, Vydac, cat no. 208TP1010) with the following gradient (Table 2).

Table 2
HPLC purification of the AMP-HDA conjugate

Time	Deionized water (%)	MeOH (%)	Flow rate (ml/min)
0	90	10	1
30	90	10	1
40	70	30	1
60	70	30	1

Table 3
HPLX purification condition of the AMP-HDA-Cy5 conjugate

Time	Deionized water (%)	MeOH (%)	Flow rate (ml/min)
0	90	10	1
25	50	50	1
40	70	30	1

6. Collect peak fraction at 53 min and lyophilize the sample.
7. Product can be stored at –80°C upon utilization.
8. In an Eppendorf tube, mix 5 μl of an AMP-HDA aqueous solution at 43 mM, with 15 μl of 1 M $NaHCO_3$ solution at pH 8.
9. Add 15 μl of Cy5 NHS ester at 10 mg/ml in extra-dry DMSO in 5 fractions separated by an interval of 10 min.
10. Vortex after each addition of Cy5 NHS ester.
11. Incubate for one more hour after the complete addition of Cy5 NHS ester.
12. Purify by reverse phase HPLC (Reverse phase C8 column, Vydac, cat no. 208TP1010) with the following gradient (Table 3).
13. Collect peak at 26 min and lyophilize the sample.
14. AMP-HDA-Cy5 can be stored at –80°C upon utilization.

3.2.2. In Vitro Transcription

Purified PCR DNA fragments are prepared as in Subheading 3.1.1. Care has to be taken in the primer design to make sure the first nucleotide incorporated will be an A.

1. In a final volume of 50 μl reaction system, mix 10 μl of 5× transcription buffer, 2 μg of DNA template, 0.5 μl of CTP (100 mM, 1 mM final), 0.5 μl of UTP (100 mM, 1 mM final), 0.5 μl of GTP (100 mM, 1 mM final), 0.125 μl of ATP (100 mM, 0.25 mM final), 10 μl of AMP-HDA-Cy5 (20 mM, 4 mM final), and 23.375 μl of 0.05% DEPC water, and then add 5 μl of T7 RNA polymerase.
2. Repeat steps 2–8 from Subheading 3.1.2.

3.3. Construction of pRNA/Folate Conjugate: Synthesis of DNA-Folate Conjugate and Annealing with pRNA

3.3.1. Synthesis of Folate-NHS and Coupling with NH_2-DNA Oligonucleotide

1. In a 100 ml round bottom flask, add folic acid (0.5 g), NHS (0.25 g) and triethylamine (0.25 ml) to 30 ml of extra-dry DMSO.
2. Agitate at 40°C until all the folic acid is dissolved then add DCC (0.5 g).
3. Continue shaking in the dark overnight at room temperature.
4. Remove insoluble byproduct by filtration of the reaction mixture through a glass wool plug.
5. Evaporate triethylamine under reduced pressure at 40°C.

Table 4
HPLC purification condition of the folate-DNA conjugate

Time	Buffer A (%)	Buffer B (%)	Flow rate (ml/min)
0	100	0	1
5	100	0	1
10	80	20	1
40	80	20	1
50	20	80	1
60	20	80	1
70	100	0	1

6. Precipitate the folate-NHS by adding 10 volumes of acetone/Et_2O (30/70).
7. Recover the product by filtration.
8. Dry the product in air for 1 h followed by drying under vacuum and store at −80°C upon utilization.
9. In a final volume of 50 μl reaction system, mix 100 μg of DNA-NH_2 oligonucleotide, 20 μl of 1 M $NaHCO_3$ solution at pH 8, then complete with 0.05% DEPC water.
10. Dissolve 5 mg of folate-NHS from step 8 in 50 μl of extra-dry DMSO.
11. Add stepwise with interval of 10 min the folate-NHS (step 10) to the DNA-NH_2 oligonucleotide solution (step 9).
12. Incubate at room temperature overnight.
13. Purify by HPLC (ProteinPak DEAE 5PW, 7.5×75 mm column) with the following gradient (Table 4).
14. Collect the peak around 18.5 min then concentrate the fraction 2–5 time.
15. Precipitate overnight at −20°C after addition of 2.5 volume of 100% ethanol.
16. Repeat steps 8 and 9 of Subheading 3.3.1.

3.3.2 .Annealing with pRNA

1. Mix equal molar amount of DNA-folate conjugate and pRNA (at least 100 ng of pRNA) in annealing buffer.
2. Incubate for 5 min in a water bath at 80°C and then let the water bath slowly cool down to room temperature.
3. Purify the pRNA/DNA-folate complex by 8% PAGE gel in TBE.
4. Repeat steps 5–8, Subheading 3.1.2.

3.4. Construction of pRNA Nanoparticles

Dimer and trimers are formed by hand-in-hand interaction of transcomplementary interlocking right and left loops. Table 1 summarizes the different matching right and left loop that can be used. Dimer and trimer preparation is described here using only a set of this transcomplementary sequences.

3.4.1. Construction of Dimeric pRNA Nanoparticles (81)

1. In TBM buffer, mix equal molar concentration of pRNA Ab′ and pRNA Ba′.
2. Purify the dimer from the monomers mixture on 8% PAGE gel in TBM (4°C, 100 V).
3. Repeat steps 5–8 from Subheading 3.1.2.

3.4.2. Construction of Trimeric pRNA Nanoparticles (37, 82, 83)

1. In TBM buffer, mix equal molar concentration of pRNA Ab′, pRNA Bc′, and pRNA Ca′.
2. Purify the trimer from the monomers mixture on 8% PAGE gel in TBM (4°C, 100 V).
3. Repeat steps 5–8 from Subheading 3.1.2.

3.5. Cross-linking with Psoralen

1. Add 1 μl of (10×)AMT solution to 0.2–10 μg of pRNA nanoparticle in 10 μl of water.
2. Spot the solution on prechilled Parafilm floating on ice water with each drop less than 10 μl.
3. Irradiate the sample with 340 nm UV with a distance of 4–5 cm for 50–60 min.
4. Load the sample in 8% polyacrylamide gel with 8 M urea in TBE (4°C, 100 V).
5. Repeat steps 5–8 from Subheading 3.1.2.

4. Notes

1. All solution should be RNase free and gloves must be worn at all times.
2. In vitro transcription of 2′-F modified pRNA with the Y639F mutant T7 RNA polymerase usually yield at least to 20% less RNA transcript as compare with the transcription of non modified pRNA.
3. Passive elution step from the gel slices described in the Subheading 3 can be replaced by an electroelution step using the Elutrap electroelution system from Whatman.
4. When Cy3 labeling is involved (Subheading 3.2), 3 M NaOAc solution are prepared at pH 6 as the Cy3 labeling is sensitive to lower pH.

Acknowledgments

This work was supported by NIH grants EB003730, GM059944, and CA151648. We also thank Dr. Randall Reif for helpful discussion and reading of the manuscript.

References

1. Niemeyer CM (2002) The developments of semisynthetic DNA-protein conjugates. Trends Biotechnol 20:395–401
2. Schmidt OG, Eberl K (2001) Nanotechnology. Thin solid films roll up into nanotubes. Nature 410:168
3. Guo P (2005) RNA nanotechnology: engineering, assembly and applications in detection, gene delivery and therapy. J Nanosci Nanotechnol 5: 1964–1982
4. Guo P, Zhang C, Chen C, Trottier M, Garver K (1998) Inter-RNA interaction of phage phi29 pRNA to form a hexameric complex for viral DNA transportation. Mol Cell 2:149–155
5. Zhang F, Lemieux S, Wu X, St.-Arnaud S, McMurray CT, Major F, Anderson D (1998) Function of hexameric RNA in packaging of bacteriophage phi29 DNA in vitro. Mol Cell 2:141–147
6. Shu D, Moll WD, Deng Z, Mao C, Guo P (2004) Bottom-up assembly of RNA arrays and superstructures as potential parts in nanotechnology. Nano Lett 4:1717–1723
7. Jaeger L, Leontis NB (2000) Tecto-RNA: one dimensional self-assembly through tertiary interactions. Angew Chem Int Ed Engl 39: 2521–2524
8. Hansma HG, Oroudjev E, Baudrey S, Jaeger L (2003) TectoRNA and 'kissing-loop' RNA: atomic force microscopy of self-assembling RNA structures. J Microsc 212:273–279
9. Cech TR, Zaug AJ, Grabowski PJ (1981) In vitro splicing of the ribosomal RNA precursor of Tetrahymena: involvement of a guanosine nucleotide in the excision of the intervening sequence. Cell 27:487–496
10. Kruger K, Grabowski PJ, Zaug AJ, Sands J, Gottschling DE, Cech TR (1982) Self-splicing RNA: autoexcision and autocyclization of the ribosomal RNA intervening sequence of Tetrahymena. Cell 31:147–157
11. Cech TR, Tanner NK, Tinoco I, Weir BR, Zuker M, Perlman PS (1983) Secondary structure of the tetrahymena ribosomal RNA intervening sequence: structural homology with fungal mitochondrial intervening sequences. Proc Natl Acad Sci USA 80:3903–3907
12. Zaug AJ, Grabowski PJ, Cech TR (1983) Autocatalytic cyclization of an excised intervening sequence RNA is a cleavage-ligation reaction. Nature 301:578–583
13. Hoeprich S, Zhou Q, Guo S, Qi G, Wang Y, Guo P (2003) Bacterial virus phi29 pRNA as a hammerhead ribozyme escort to destroy hepatitis B virus. Gene Ther 10:1258–1267
14. Liu H, Guo S, Roll R, Li J, Diao Z, Shao N, Riley MR, Cole AM, Robinson JP, Snead NM, Shen G, Guo P (2007) Phi29 pRNA vector for efficient escort of hammerhead ribozyme targeting survivin in multiple cancer cells. Cancer Biol Ther 6:697–704
15. Guo P (2010) The emerging field of RNA nanotechnology. Nat Nanotechnol 5:833–842
16. Bouvet P (2001) Determination of nucleic acid recognition sequences by SELEX. Methods Mol Biol 148:603–610
17. Ciesiolka J, Gorski J, Yarus M (1995) Selection of an RNA domain that binds Zn^{2+}. RNA 1:538–550
18. Clark S, Remcho V (2002) Aptamers as analytical reagents. Electrophoresis 23:1335–1340
19. Kraus E, James W, Barclay AN (1998) Cutting edge: novel RNA ligands able to bind CD4 antigen and inhibit CD4+ T lymphocyte function. J Immunol 160:5209–5212
20. Shu D, Guo P (2003) A viral RNA that binds ATP and contains an motif similar to an ATP-binding aptamer from SELEX. J Biol Chem 278:7119–7125
21. Gold L (1995) The SELEX process: a surprising source of therapeutic and diagnostic compounds. Harvey Lect 91:47–57
22. Ellington AD, Szostak JW (1990) *In vitro* selection of RNA molecules that bind specific ligands. Nature 346:818–822
23. Tuerk C, Gold L (1990) Systematic evolution of ligands by exponential enrichment: RNA ligands to bacteriophage T4 DNA ploymerase. Science 249:505–510
24. Khaled A, Guo S, Li F, Guo P (2005) Controllable self-assembly of nanoparticles for specific delivery of multiple therapeutic molecules to cancer cells using RNA nanotechnology. Nano Lett 5:1797–1808

25. McNamara JO, Andrechek ER, Wang Y, Viles KD, Rempel RE, Gilboa E, Sullenger BA, Giangrande PH (2006) Cell type-specific delivery of siRNAs with aptamer-siRNA chimeras. Nat Biotechnol 24:1005–1015
26. Dassie JP, Liu XY, Thomas GS, Whitaker RM, Thiel KW, Stockdale KR, Meyerholz DK, McCaffrey AP, McNamara JO, Giangrande PH (2009) Systemic administration of optimized aptamer-siRNA chimeras promotes regression of PSMA-expressing tumors. Nat Biotechnol 27:839–849
27. Zhou J, Swiderski P, Li H, Zhang J, Neff CP, Akkina R, Rossi JJ (2009) Selection, characterization and application of new RNA HIV gp 120 aptamers for facile delivery of Dicer substrate siRNAs into HIV infected cells. Nucleic Acids Res 37:3094–3109
28. Zhou J, Li H, Zaia J, Rossi JJ (2008) Novel dual inhibitory function aptamer-siRNA delivery system for HIV-1 therapy. Mol Ther 16: 1481–1489
29. Cerchia L, de Franciscis V (2010) Targeting cancer cells with nucleic acid aptamers. Trends Biotechnol 28:517–525
30. Fire A, Xu S, Montgomery MK, Kostas SA, Driver SE, Mello CC (1998) Potent and specific genetic interference by double-stranded RNA in *Caenorhabditis elegans*. Nature 391:806–811
31. Ghildiyal M, Zamore PD (2009) Small silencing RNAs: an expanding universe. Nat Rev Genet 10:94–108
32. Guo P, Coban O, Snead NM, Trebley J, Hoeprich S, Guo S, Shu Y (2010) Engineering RNA for targeted siRNA delivery and medical application. Adv Drug Deliv Rev 62:650–666
33. Li H, Li WX, Ding SW (2002) Induction and suppression of RNA silencing by an animal virus. Science 296:1319–1321
34. Brummelkamp TR, Bernards R, Agami R (2002) A system for stable expression of short interfering RNAs in mammalian cells. Science 296:550–553
35. Jacque JM, Triques K, Stevenson M (2002) Modulation of HIV-1 replication by RNA interference. Nature 418:435–438
36. Varambally S, Dhanasekaran SM, Zhou M, Barrette TR, Kumar-Sinha C, Sanda MG, Ghosh D, Pienta KJ, Sewalt RG, Otte AP, Rubin MA, Chinnaiyan AM (2002) The polycomb group protein EZH2 is involved in progression of prostate cancer. Nature 419:624–629
37. Carmichael GG (2002) Medicine: silencing viruses with RNA. Nature 418:379–380
38. Sarver NA, Cantin EM, Chang PS, Zaia JA, Ladne PA, Stephens DA, Rossi JJ (1990) Ribozymes as potential anti-HIV-1 therapeutic agents. Science 247:1222–1225
39. Chowrira BM, Berzal-Herranz A, Burke JM (1991) Novel guanosine requirement for catalysis by the hairpin ribozyme. Nature 354:320–322
40. Forster AC, Symons RH (1987) Self-cleavage of virusoid RNA is performed by the proposed 55- nucleotide active site. Cell 50:9–16
41. Nava Sarver N, Cantin EM, Chang PS, Zaia JA, Ladne PA, Stephens DA, Rossi JJ (1990) Ribozymes as potential anti-HIV-1 therapeutic agents. Science 24:1222–1225
42. Guo S, Huang F, Guo P (2006) Construction of folate-conjugated pRNA of bacteriophage phi29 DNA packaging motor for delivery of chimeric siRNA to nasopharyngeal carcinoma cells. Gene Ther 13:814–820
43. Coleman J, Hirashima A, Inocuchi Y, Green PJ, Inouye M (1985) A novel immune system against bacteriophage infection using complementary RNA (micRNA). Nature 315:601–603
44. Knecht DA, Loomis WF (1987) Antisense RNA inactivation of myosin heavy chain gene expression in Dictyostelium discoideum. Science 236:1081–1086
45. Liu J, Guo S, Cinier M, Shlyakhtenko L, Shu Y, Chen C, Shen G, Guo P (2010) Fabrication of stable and RNase—resistant RNA nanoparticles active in gearing the nanomotors for viral DNA packaging. ACS Nano 5:237–246
46. Shu Y, Cinier M, Shu D, Guo P (2011) Assembly of multifunctional phi29 pRNA nanoparticles for specific delivery of siRNA and other therapeutics to targeted cells. Methods 54(2):204–214
47. Abdelmawla S, Guo S, Zhang L, Pulukuri S, Patankar P, Conley P, Trebley J, Guo P, Li QX (2011) Pharmacological characterization of chemically synthesized monomeric pRNA nanoparticles for systemic delivery. Mol Ther 19(7):1312–1322
48. Shu Y, Cinier M, Fox SR, Ben-Johnathan N, Guo P (2011) Assembly of therapeutic pRNA-siRNA nanoparticles using bipartite approach. Mol Ther 19(7):1304
49. Guo P, Erickson S, Anderson D (1987) A small viral RNA is required for *in vitro* packaging of bacteriophage phi29 DNA. Science 236: 690–694
50. Shu D, Zhang H, Jin J, Guo P (2007) Counting of six pRNAs of phi29 DNA-packaging motor with customized single molecule dual-view system. EMBO J 26:527–537
51. Chen C, Zhang C, Guo P (1999) Sequence requirement for hand-in-hand interaction in formation of pRNA dimers and hexamers to gear phi29 DNA translocation motor. RNA 5:805–818
52. Guo S, Tschammer N, Mohammed S, Guo P (2005) Specific delivery of therapeutic RNAs to

cancer cells via the dimerization mechanism of phi29 motor pRNA. Hum Gene Ther 16: 1097–1109

53. Glotzer SC (2004) Materials science. Some assembly required. Science 306:419–420
54. Gates BD, Xu Q, Stewart M, Ryan D, Willson CG, Whitesides GM (2005) New approaches to nanofabrication: molding, printing, and other techniques. Chem Rev 105:1171–1196
55. Tarapore P, Shu Y, Guo P, Ho SM (2010) Application of Phi29 motor pRNA for targeted therapeutic delivery of siRNA silencing metallothionein-IIA and survivin in ovarian cancers. Mol Ther 19:386–394
56. Zhang HM, Su Y, Guo S, Yuan J, Lim T, Liu J, Guo P, Yang D (2009) Targeted delivery of anti-coxsackievirus siRNAs using ligand-conjugated packaging RNAs. Antiviral Res 83:307–316
57. Shu Y, Shu D, Diao Z, Shen G, Guo P (2009) Fabrication of polyvalent therapeutic RNA nanoparticles for specific delivery of siRNA, ribozyme and drugs to targeted cells for cancer therapy. IEEE/NIH Life Science Systems and Applications Workshop, pp 9–12
58. Rusckowski M, Qu T, Roskey A, Agrawal S (2000) Biodistribution and metabolism of a mixed backbone oligonucleotide (GEM 231) following single and multiple dose administration in mice. Antisense Nucleic Acid Drug Dev 10:333–345
59. Kawasaki AM, Casper MD, Freier SM, Lesnik EA, Zounes MC, Cummins LL, Gonzalez C, Cook PD (1993) Uniformly modified 2′-deoxy-2′-fluoro phosphorothioate oligonucleotides as nuclease-resistant antisense compounds with high affinity and specificity for RNA targets. J Med Chem 36:831–841
60. Pieken WA, Olsen DB, Benseler F, Aurup H, Eckstein F (1991) Kinetic characterization of ribonuclease-resistant 2′-modified hammerhead ribozymes. Science 253:314–317
61. Jaeger L, Verzemnieks EJ, Geary C (2009) The UA_handle: a versatile submotif in stable RNA architectures. Nucleic Acids Res 37:215–230
62. Huang Y, Eckstein F, Padilla R, Sousa R (1997) Mechanism of ribose 2′-group discrimination by an RNA polymerase. Biochemistry 36: 8231–8242
63. Padilla R, Sousa R (1999) Efficient synthesis of nucleic acids heavily modified with non-canonical ribose 2′-groups using a mutantT7 RNA polymerase (RNAP). Nucleic Acids Res 27: 1561–1563
64. Sousa R, Padilla R (1995) A mutant T7 RNA polymerase as a DNA polymerase. EMBO J 14:4609–4621
65. Padilla R, Sousa R (2002) A Y639F/H784A T7 RNA polymerase double mutant displays superior properties for synthesizing RNAs with non-canonical NTPs. Nucleic Acids Res 30:e138
66. Braasch DA, Jensen S, Liu Y, Kaur K, Arar K, White MA, Corey DR (2003) RNA interference in mammalian cells by chemically-modified RNA. Biochemistry 42:7967–7975
67. Harborth J, Elbashir SM, Vandenburgh K, Manninga H, Scaringe SA, Weber K, Tuschl T (2003) Sequence, chemical, and structural variation of small interfering RNAs and short hairpin RNAs and the effect on mammalian gene silencing. Antisense Nucleic Acid Drug Dev 13:83–105
68. Elmen J, Thonberg H, Ljungberg K, Frieden M, Westergaard M, Xu Y, Wahren B, Liang Z, Orum H, Koch T, Wahlestedt C (2005) Locked nucleic acid (LNA) mediated improvements in siRNA stability and functionality. Nucleic Acids Res 33:439–447
69. Layzer JM, McCaffrey AP, Tanner AK, Huang Z, Kay MA, Sullenger BA (2004) In vivo activity of nuclease-resistant siRNAs. RNA 10:766–771
70. Huang F, Wang G, Coleman T, Li N (2003) Synthesis of adenosine derivatives as transcription initiators and preparation of 5′ fluorescein- and biotin-labeled RNA through one-step *in vitro* transcription. RNA 9:1562–1570
71. Li N, Yu C, Huang F (2005) Novel cyanine-AMP conjugates for efficient 5′ RNA fluorescent labeling by one-step transcription and replacement of [gamma-32P]ATP in RNA structural investigation. Nucleic Acids Res 33:e37
72. Huang F, He J, Zhang Y, Guo Y (2008) Synthesis of biotin-AMP conjugate for 5′ biotin labeling of RNA through one-step in vitro transcription. Nat Protoc 3:1848–1861
73. Shu D, Zhang H, Petrenko R, Meller J, Guo P (2010) Dual-channel single-molecule fluorescence resonance energy transfer to establish distance parameters for RNA nanoparticles. ACS Nano 4:6843–6853
74. Zhang CL, Trottier M, Guo PX (1995) Circularly permuted viral pRNA active and specific in the packaging of bacteriophage f29 DNA. Virology 207:442–451
75. Zhang CL, Lee C-S, Guo P (1994) The proximate 5′ and 3′ ends of the 120-base viral RNA (pRNA) are crucial for the packaging of bacteriophage f29 DNA. Virology 201:77–85
76. Zhang L, Sun L, Cui Z, Gottlieb RL, Zhang B (2001) 5′-sulfhydryl-modified RNA: initiator synthesis, in vitro transcription, and enzymatic incorporation. Bioconjug Chem 12:939–948

77. Milligan JF, Groebe DR, Witherell GW, Uhlenbeck OC (1987) Oligoribonucleotide synthesis using T7 RNA polymerase and synthetic DNA templates. Nucleic Acids Res 15:8783–8798
78. Bruce AG, Uhlenbeck OC (1978) Reactions at the termini of tRNA with T4 RNA ligase. Nucleic Acids Res 5:3665–3677
79. Garver K, Guo P (2000) Mapping the inter-RNA interaction of phage phi29 by site-specific photoaffinity crosslinking. J Biol Chem 275:2817–2824
80. Huang F (2003) Efficient incorporation of CoA, NAD and FAD into RNA by in vitro transcription. Nucleic Acids Res 31:e8
81. Pan T, Gutell RR, Uhlenbeck OC (1991) Folding of circularly permuted transfer RNAs. Science 254:1361–1364
82. Nolan JM, Burke DH, Pace NR (1993) Circularly Permuted tRNAs as Specific Photoaffinity Probes of Ribonuclease P RNA Structure. Science 261:762–765
83. Elbashir SM, Harborth J, Lendeckel W, Yalcin A, Weber K, Tuschl T (2001) Duplexes of 21-nucleotide RNAs mediate RNA interference in cultured mammalian cells. Nature 411:494–498
84. Chen C, Guo P (1997) Magnesium-induced conformational change of packaging RNA for procapsid recognition and binding during phage phi29 DNA encapsidation. J Virol 71:495–500
85. Wassarman DA (1993) Psoralen crosslinking of small RNAs *in vitro*. Mol Biol Rep 17:143–151
86. Tyc K, Steitz JA (1992) A new interaction between the mouse 5′ external transcribed spacer of pre-rRNA and U3 snRNA detected by psoralen crosslinking. Nucleic Acids Res 20:5375–5382
87. Hui CF, Cantor CR (1985) Mapping the location of psoralen crosslinks on RNA by mung bean nuclease sensitivity of RNA-DNA hybrids. Proc Natl Acad Sci USA 82:1381–1385
88. Shu D, Huang L, Hoeprich S, Guo P (2003) Construction of phi29 DNA-packaging RNA (pRNA) monomers, dimers and trimers with variable sizes and shapes as potential parts for nanodevices. J Nanosci Nanotechnol 3:295–302
89. Chen C, Sheng S, Shao Z, Guo P (2000) A dimer as a building block in assembling RNA: a hexamer that gears bacterial virus phi29 DNA-translocating machinery. J Biol Chem 275: 17510–17516

Chapter 17

Mouse Models for Tumor Metastasis

Shengyu Yang, J. Jillian Zhang, and Xin-Yun Huang

Abstract

Tumor metastasis is the main cause of death of cancer patients. Here we describe two mouse models for investigating tumor metastasis. In the first spontaneous metastasis mouse model, 4T1 mouse breast tumor cells are injected into the mammary gland of host mice and the metastasis of 4T1 tumor cells into the lung are examined with a colonogenic assay. In the second experimental metastasis mouse model, luciferase-labeled MDA-MB-231 human breast tumor cells are injected into the tail vein of NOD-SCID immunodeficient mice and the colonization of MDA-MB-231 tumor cells in the lung are monitored using noninvasive bioluminescence imaging.

Key words: Tumwor metastasis, 4T1 mouse breast tumor cells, MDA-MB-231 human breast tumor cells, Allograft, Xenograft

1. Introduction

Metastasis is the multistep process wherein a primary tumor spreads from its initial site to secondary tissues/organs (1, 2). Despite the significant improvement in both diagnostic and therapeutic modalities for the treatment of cancer patients, tumor metastasis is still the major cause of mortality being responsible for ~90% of all cancer deaths (3, 4). Therefore, development of therapeutic agents that prevent tumor metastasis is very desirable.

To metastasize, cancer cells have to succeed in invasion, intravasation, survival in the circulation, extravasation, and proliferation within the distant parenchyma (5–7). In our laboratory, we have used two mouse models for tumor metastasis (8–10). The first model is an orthotopic allograft model (also called the spontaneous tumor metastasis model). In this model, 4T1 mouse mammary tumor cells are injected into the second mammary gland of syngenetic Balb/c mice and the spontaneous metastasis to the lung will

Yi Zheng (ed.), *Rational Drug Design: Methods and Protocols*, Methods in Molecular Biology, vol. 928,
DOI 10.1007/978-1-62703-008-3_17, © Springer Science+Business Media New York 2012

be analyzed with a colonogenic assay (11, 12). The second model is an experimental metastasis model. In this model, luciferase-labeled MDA-MB-231 human breast tumor cells are injected into the tail vein of NOD-SCID immunodeficient mice and the presence of tumor cells in mice will be detected through the noninvasive bioluminescence imaging (13, 14).

2. Materials

1. 4T1 growth medium: RPMI 1640 medium supplemented with 10% FBS and 1% antibiotic-antimyotic.
2. 100× Antibiotic-antimyotic.
3. 0.25% Trypsin/1 mM EDTA.
4. Phosphate buffered saline.
5. Collagenase solution: 2 mg/ml collagenase, Type IV, 0.01% DNase I in PBS. Filter the solution through a 0.2 μm filter to sterilize.
6. 6-Thioguanine: Dissolve 40 mg 6-thioguanine with 4 ml 0.1 N NaOH and filter the solution with 0.2 μm filter to sterilize.
7. Selection Medium: 4T1 growth medium supplemented with 10 μg/ml 6-thioguanine.
8. Crystal violet staining solution.
9. 70-μm Cell strainer.
10. 60-cm Tissue culture dishes.
11. Dissecting scissors, curved scissors, and dissecting tissue forceps.
12. 4T1 mouse breast tumor cells and MDA-MB-231 human breast tumor cells (ATCC).
13. 6–8-weeks-old female BALB/c mice (Charles River) and 4–6 weeks old female NOD-SCID mice.
14. D-Luciferin: 10 mg/ml, dissolved in PBS and filter to sterilize.
15. Isoflurane, isoflurane vaporizer, induction chamber, oxygen tank, and gas regulator.
16. PEI stock solution: Dissolve polyethylenimine powder to a concentration of 1 mg/ml in water and adjust pH to 7.0 with 5 M HCl. Filter sterilizes and freeze aliquots at −80°C. If PEI precipitates after freeze and thaw just resuspend well before used for transfection. The precipitation does not affect the transfection efficiency.
17. DMEM with or without 10% FBS.
18. Polybrene: 10 mg/ml, dissolved in water; filter to sterilize.

3. Methods

3.1. 4T1 Cell Allograft Metastasis Mouse Model

3.1.1. 4T1 Cell Culture and Harvest

1. Culture 4T1 mouse mammary tumor cells in growth medium in a 37°C, 5% CO_2 tissue culture incubator. 4T1 cells double about every 12 h and should be split 1:5 or 1:10 every 2–3 days (see Note 1).
2. Discard culture medium and wash 4T1 cells in a 10 cm tissue culture plate with 5 ml PBS. Add 2 ml 1× Trypsin/EDTA solution to the plate and swirl to cover the entire plate. Incubate at room temperature for about 2 min and tap the side of the plate occasionally (see Note 2).
3. When the cells detach from the plate, stop the trypsinization by adding 2 ml growth medium to the plate and suspend the cells by pipetting up and down a couple times.
4. Spin down the cells at 500 rpm, room temperature in a bench-top centrifuge. Wash the cells once with 5 ml RPMI starvation medium and resuspend the cells in 1 ml RPMI medium. Determine the cell number with a hemocytometer. Dilute the cells to 5×10^6/ml with RPMI starvation medium.

3.1.2. Orthotopic Allograft

1. Shave female Balb/c mice (4–6 weeks old) with a hair clipper around the second mammary gland.
2. Load a 1 ml tuberculin syringe with cells. Attach a 26 gauge ½ in. needle to the syringe; align the bevel of the needle toward the metric numbers on the syringe. Carefully eliminate air bubbles in the needle and syringe (see Note 3).
3. Restrain a 6–8 weeks old female Balb/c mouse with one hand and inject 100 μl 4T1 cell (5×10^5 cells) subcutaneously into the second mammary gland. Palpable primary tumors usually develop within 1 week (see Note 4).
4. Administer mice with antimetastasis agents 1 week after the tumor cell injection. Antimetastasis agents are typically diluted into 100 μl PBS and administered intraperitoneally on a daily basis, although other route of administration or doses could also be used. 100 μl PBS or other vector could be administered as control.

3.1.3. Colonogenic Assay

1. Euthanize the mice 4 weeks after the tumor cell injection or when they are moribund. Spray the mice with 70% ethanol so that the fur on the chest is wet. Open up the chest with dissecting scissors and forceps, and retrieve the lung.
2. Rinse the lung with 2.5 ml PBS and then transfer the lung to another 2.5 ml PBS in a 60-mm tissue culture plate. Mince the lung to pieces as small as possible with curved scissors. Transfer the lung tissue suspension to a 15-ml Falcon tube containing

2.5 ml of collagenase/DNase I solution using a 5 ml tissue culture pipette. Thoroughly minced lung tissue should not clog the pipette.

3. Incubate the mixture in a 37°C shaker and shake the tube at 250 rpm for 2 h.
4. Gently pipette the suspension a couple of times to break up the chunks and filter the solution through a 70 μm cell strainer.
5. Spin down the cells, wash the pellet with 10 ml PBS once and resuspend the cells in 10 ml Selection Medium. Make four 1/10 serial dilutions by removing 1 ml from higher concentration cell suspension and mixing with 9 ml Selection Medium. Culture the cells in a 7% CO_2 incubator at 37°C for 10–14 days to allow the formation of 4T1 cell colonies (see Note 5).
6. Discard the Selection Medium and fix the cells by adding 5 ml methanol to each plate and incubate at room temperature for 5 min. Rehydrate the cells by washing with 5 ml distilled water and stain the cells with 2 ml crystal violet staining solution. Swirl to cover the entire plate with staining solution. Wash the cells a couple of times with distilled water and count the number of colonies in the dilution with about 100 colonies.

3.2. Human Breast Tumor Cell Xenograft Mouse Model

3.2.1. Labeling Human Breast Tumor Cells with Luciferase Using Retroviruses

1. Culture 293T cells in DMEM medium supplemented with 10% FBS and antibiotic-antimyotic agents.
2. Split one confluent plate of 293T cells 1:3 12–16 h before transfection. Feed the cells with fresh growth medium 2–3 h prior to the transfection. The cells should be 60–70% confluent at the time of transfection.
3. We used a triple-fusion protein reporter retrovirus construct encoding herpes simplex virus thymidine kinase 1, green fluorescent protein (GFP) and firefly luciferase (TGL) (9, 13). However, any retroviral vector encoding fire fly luciferase can also be used. Mix luciferase retroviral construct with VSV-G (e.g., Addgene plasmid 12259) and gag-pol (e.g., Addgene plasmid 14487), 5 μg each and dilute the plasmid with 500 μl Opti-MEM medium. Dilute 45 μl PEI stock (1 mg/ml) with 500 μl Opti-MEM and incubate at room temperature for 5 min (see Note 6).
4. Mix the DNA solution and the PEI solution and vortex vigorously for 10 s. Incubate the mixture at room temperature for 15 min (see Note 7).
5. Add the mixture, drop by drop, to a 10-cm dish of 293T cells. Return the cell to the 37°C 5% CO_2 incubator.
6. 12–16 h later discard the medium and feed the cells with fresh growth medium.

7. Harvest retroviruses in the medium 48 h and 72 h after transfection.
8. Centrifuge the medium containing retroviruses at 4,000 × *g* for 30 min at 4°C to remove cell debris. Carefully transfer the supernatant to a transparent centrifuge tube.
9. Centrifuge the supernatant at 48,000 × *g* for 2 h at 4°C. The virus precipitate appears as a small white or yellow pellet on the wall of the centrifuge tube. Carefully remove the supernatant without disturbing the pellet. Resuspend the viruses in growth medium. Typically 1 ml growth medium is used for every 10 ml of supernatant. The retroviruses can be snap-frozen with liquid nitrogen and stored at –80°C or used immediately for infection (see Note 8).
10. To label MDA-MB-231 cells with luciferase, 1 ml concentrated viruses encoding luciferase are added to cells in a 10-cm culture dish (50–70% confluent). Add polybrene to the final concentration of 8 μg/ml.
11. The medium can be changed 12 h post-infection and the cells can be assayed for luciferase activity 72 h post-infection.

3.2.2. Tail Vein Injection of MDA-MB-231 Cells

1. Culture MDA-MB-231 human breast cancer cells stably expressing firefly luciferase in a 37°C, 5% CO_2 tissue culture incubator.
2. Trypsinize to detach cells from the culture plate and determine cell number with a hemocytometer. Dilute the cells with DMEM medium (without FBS) to 5×10^6/ml.
3. Use a heating lamp to dilate the tail vein of female NOD-SCID mice (4–6 weeks old). More than one animal can be heated together in a plastic cage, with a heating lamp above them. Proper dilation of the tail vein is crucial to the success of tail vein injection (see Note 9).
4. Transfer a mouse to a restraining device; properly secure the mouse with the tail exposed.
5. Load a 1 ml tuberculin syringe with luciferase-labeled tumor cells. Attach a 26 gauge ½ in. needle to the syringe; align the bevel of the needle toward the metric numbers on the syringe. Carefully eliminate air bubbles in the needle and syringe. Inject 100 μl luciferase-labeled cells into one of the lateral tail veins.

3.2.3. Noninvasive Bioluminescence Imaging

1. Check the level of isofluorane in the vaporizer and gas level in the oxygen tank to ensure adequate amount of isofluorane and oxygen for the duration of the procedure. Make sure that the system is set to flow to the induction chamber and/or IVIS imaging system.

2. Turn on the supply gas and adjust the flow meter to 500–1,000 ml/min.
3. Place mice in the induction chamber and seal the chamber. Adjust the vaporizer to 2.5–3% and monitor the mice until recumbent. Lift one end of the induction chamber to roll the mice over to ensure that the animals are properly anesthetized.
4. Inject 100 μl d-Luciferin (15 mg/ml) via i.p. into each mouse. Transfer the mice to the IVIS imaging machine. Align the noses of the mice with the nosecones to ensure that the mice stay immobile during the imaging process.
5. Record the luminescence image with the IVIS imaging machine.
6. Image the mice on day 0 (within 2 h after the xenograft), on day 1, and on a weekly basis thereafter.
7. Properly label the mice so that each individual mouse can be identified in the imaging process. Measure the luminescence photon flux signal by selecting a ROI (region of interest) over the thoracic region of the mice. Calculate the normalized photon flux using the following formula:

$$P_n = 100 \times (P_t / P_0),$$

where P_n is the normalized photon flux, P_t is the photon flux of a given date, and P_0 is the photon flux signal of day 0 (see Note 10).

4. Notes

1. Overgrowth reduces the viability of 4T1 cells and MDA-MB-231 cells. Therefore, precaution should be taken to avoid overgrowth. This is especially important for 4T1 cells, which is a rapidly proliferating cell line that doubles about every 12 h.
2. When harvesting the tumor cells for xenograft/allograft, cells should not be overtrypsinized. Overtrypsinization reduces the invasiveness of tumor cells.
3. It takes some practice to properly inject 4T1 cells subcutaneously. Proper injection is the key to minimize variation among individual animals. When properly injected, the tumor cell suspension looks like a “bubble” underneath the skin.
4. To avoid contamination, mice need to be thoroughly sprayed with 70% ethanol before harvesting the lungs for colonogenic assays. The forceps and scissors need to be soaked in 70%

ethanol for at least 30 min before starting the experiment, and washed with 70% ethanol before used on a different animal.

5. Do not disturb the cell culture plates during the colonogenic assay. It would be best to culture the plates in a separate CO_2 incubator committed for this assay. Frequent opening and closing of incubator doors and moving the plates during the 10–14 day incubation may result in the formation of secondary colonies and artificially increase colony numbers.
6. The VSV-G pseudotyped retrovirus could be prepared using packaging cell lines stably expressing VSV-G and gag-pol, or by co-transfection of 293T cells with plasmids encoding VSV-G and gag-pol. We find that retroviruses produced with co-transfection approach generally have much higher titers.
7. PEI transfection is an affordable and high efficient substitute for other lipid based transfection reagents such as Lipofectamine 2000. PEI transfection routinely gives us high transfection efficiency close to 100%, and is much more consistent and reproducible than calcium phosphate method.
8. Biosafety level 2 standard should be taken when handling VSV-G pseudotyped retroviruses. All the utensil and disposable plastics must be treated with concentrated bleach.
9. The proper dilation of tail vein with heating lamp is the key to tail vein injection. When properly dilated, the veins look bulged and the tail feels warm.
10. The mice need to be properly labeled before the first bioluminescence imaging. The photon flux derived from the first imaging (day 0) should be used to normalize the luminescence signal from subsequent imaging.

Acknowledgment

This work was supported by NIH grant R01CA136837 to Xin-Yun Huang and a Career Development Award from Donald A. Adam Comprehensive Melanoma Research Center to Shengyu Yang.

References

1. Weiss L (2000) Metastasis of cancer: a conceptual history from antiquity to the 1990s. Cancer Metastasis Rev 19:I–XI, 193–383
2. Fidler IJ (2003) The pathogenesis of cancer metastasis: the 'seed and soil' hypothesis revisited. Nat Rev Cancer 3:453–458
3. Hanahan D, Weinberg RA (2000) The hallmarks of cancer. Cell 100:57–70
4. Christofori G (2006) New signals from the invasive front. Nature 441:444–450
5. Partin AW, Schoeniger JS, Mohler JL, Coffey DS (1989) Fourier analysis of cell motility: correlation of motility with metastatic potential. Proc Natl Acad Sci USA 86:1254–1258
6. Aznavoorian S, Murphy AN, Stetler-Stevenson WG, Liotta LA (1993) Molecular aspects of

tumor cell invasion and metastasis. Cancer 71:1368–1383

7. Condeelis J, Singer RH, Segall JE (2005) The great escape: when cancer cells hijack the genes for chemotaxis and motility. Annu Rev Cell Dev Biol 21:695–718
8. Shan D, Chen L, Njardarson JT, Gaul C, Ma X, Danishefsky SJ, Huang XY (2005) Synthetic analogues of migrastatin that inhibit mammary tumor metastasis in mice. Proc Natl Acad Sci USA 102:3772–3776
9. Yang S, Zhang JJ, Huang XY (2009) Orai1 and STIM1 are critical for breast tumor cell migration and metastasis. Cancer Cell 15:124–134
10. Chen L, Yang S, Jakoncic J, Zhang JJ, Huang XY (2010) Migrastatin analogues target fascin to block tumour metastasis. Nature 464: 1062–1066
11. Mitsuhashi M, Liu J, Cao S, Shi X, Ma X (2004) Regulation of interleukin-12 gene expression and its anti-tumor activities by prostaglandin E2 derived from mammary carcinomas. J Leukoc Biol 76:322–332
12. Pulaski BA, Ostrand-Rosenberg S (1998) Reduction of established spontaneous mammary carcinoma metastases following immunotherapy with major histocompatibility complex class II and B7.1 cell-based tumor vaccines. Cancer Res 58:1486–1493
13. Minn AJ, Kang Y, Serganova I, Gupta GP, Giri DD, Doubrovin M, Ponomarev V, Gerald WL, Blasberg R, Massague J (2005) Distinct organ-specific metastatic potential of individual breast cancer cells and primary tumors. J Clin Invest 115:44–55
14. Minn AJ, Gupta GP, Siegel PM, Bos PD, Shu W, Giri DD, Viale A, Olshen AB, Gerald WL, Massague J (2005) Genes that mediate breast cancer metastasis to lung. Nature 436: 518–524

Index

A

Acquired drug resistance 176
Activation loop 153, 154
Affinity purification 143–150
AGC protein kinase 134
Allograft 221, 223–224, 226
Allosteric
 inhibition 140
 inhibitor 67–78, 134
 regulation 68, 134
 site 134, 138, 140, 153
Alpha helical mimetic molecules 128
Antagomir 185–195
Antibody–carbohydrate interactions 42–45
ATP-binding site 133–135, 138, 140, 143, 154
AutoDock 3, 4, 7–10, 12–14
Autoinhibition 69, 72

B

Bacteriophage phi29 198
Bottom-up assembly 199

C

Cancer 17–19, 29, 40, 53, 70, 81, 161, 162, 175, 186, 198, 200, 201, 206, 207, 221, 225
Cdc42 ....31, 67, 68, 70–76, 82, 94
Cell-type specific delivery 198
Chemicals 3, 17–19, 21, 24, 25, 30, 33, 42, 68, 81–95, 116, 136, 138, 144, 145, 155, 176, 199–204, 206, 212
Clustering 11, 13, 42
Conformational change 2, 5–7, 26, 37, 133–141

D

Database search 3, 18, 19, 24
DFG-out 144, 153, 154, 156
 conformation 144, 153, 154
Dimer disruptor 120, 121, 128, 130
Dissociation constant 21, 39, 46, 156, 158
Docking 1–14, 17, 18, 20, 22, 24–26, 31, 32, 41, 46, 51, 133
Drug
 delivery 200
 discovery 1, 18, 39

F

Flexible docking 2, 6
Fluorescence
 polarization 54, 56, 61
 tagged 54, 56, 58–59, 170
Fluorescent
 GEF assay 99, 101, 116
 probe 154–158
Fluorogenic enzyme assay 121

G

Ganglioside GD2 ....40, 41, 43, 47–51
GTP-binding protein 115

H

HHV. *See* Human herpesvirus protease (HHV)
High throughput screening (HTS) 18, 54, 59–60, 120, 121, 124, 162, 170–172
Human herpesvirus protease (HHV) 119–121

I

Inactive conformation 143–150, 153–159
Inhibitor 5, 17–27, 29–37, 53–64, 67–78, 81–95, 97–117, 119–130, 133, 134, 143–145, 148, 153, 154, 162, 166, 167, 169, 170, 172, 174, 176, 177
 screening 67, 82, 89–90, 97–118

K

Kaposi's sarcoma-associated herpesvirus (KSHV) 120, 121, 123–126, 128, 129
Kinase 67–78, 133–141, 143–150, 153–159, 163, 166, 173, 175, 176, 181, 183, 224
 inhibitor 67, 133, 153, 176
KSHV. *See* Kaposi's sarcoma-associated herpesvirus (KSHV)

M

MDA-MB-231 human breast tumor cells (ATCC) 222, 224–225

Yi Zheng (ed.), *Rational Drug Design: Methods and Protocols*, Methods in Molecular Biology, vol. 928, DOI 10.1007/978-1-62703-008-3, © Springer Science+Business Media New York 2012

MicroRNA ... 185–195
Molecular modeling ... 1, 5, 40–42, 46–47
Murine bone marrow cells ... 185–195

N

Nanobiotechnology ... 197, 200, 202, 203
Nanomotors ... 202
^{15}N HSQC ... 21
Nuclear magnetic resonance (NMR) ... 1, 17–27, 39–51, 121, 130
  structure calculations ... 45, 50

P

Pak kinase ... 68, 71
PCR-based mutagenesis ... 98, 99, 101, 106–107
PDK1 ... 133–141
Peptide
  aptamers ... 97–117
  inhibitors ... 115
  mimetics ... 128
Peptide–carbohydrate interaction ... 41
Peripheral binding site ... 55, 56
Polyview-MM ... 5, 12, 13
pRNA nanoparticle ... 197–216
Protein
  conformation ... 134, 138, 153–159
  kinase ... 133–141, 143–150, 153–159, 175
  ligands ... 5, 13, 20, 22, 46
  structure ... 19, 128
  tyrosine phosphatases (PTPs) ... 53–64
PTP inhibitor ... 53–64

R

Rational drug design ... 30, 40, 153
Re-ranking ... 2, 17, 24
Retroviral-screening ... 182
RhoA ... 29–36, 82, 94, 98, 101, 113
RhoGEF. *See* Rho guanine nucleotide factor (RhoGEF)
Rho GTPases ... 29, 30, 81, 82, 94
Rho guanine nucleotide factor (RhoGEF) ... 30, 80, 82, 98, 101
Rho signaling ... 81, 82
RNA interference (RNAi) ... 198
RNA nanotechnology ... 202

S

SAR. *See* Structure–activity relationship (SAR)
Saturation transfer difference NMR ... 39
Screening ... 1–14, 17–27, 30, 32, 33, 54, 56, 59–61, 67–69, 71, 72, 75–76, 82, 88–92, 97–130, 162, 170–172, 175–183
Signaling ... 17–27, 29, 30, 81, 82, 98, 175
Small molecule
  docking ... 3
  inhibitors ... 30, 31, 36, 53, 119–130, 143
Structure ... 1, 2, 5–9, 11, 13, 14, 17–26, 30, 39–42, 44–46, 49–51, 55, 56, 120, 122, 128, 140, 144, 153–155, 178, 198–199, 202, 204, 206, 207
Structure–activity relationship (SAR) ... 18, 25, 153

T

Targeting ... 17–27, 29–37, 40, 55, 67, 98, 101, 120, 133–135, 143, 200, 205, 206
4T1 mouse breast tumor cells ... 222
Transferred NOE ... 40, 48–49
Tumor metastasis ... 40, 221–227
Two-hybrid ... 81–94, 98–100, 102–104, 110–112
Tyrosine kinase inhibitors ... 176

V

Viral
  assembly ... 197, 199, 200
  DNA packaging motor ... 198
Virtual ligand screening ... 17–27
Virtual screening ... 1–14, 18, 19, 24, 26, 30, 32, 33
Visualization ... 1, 5, 6, 11, 13, 150

X

Xenograft ... 206, 224–226

Y

Yeast two-hybrid ... 81–94, 99, 100, 103–104, 110–112

Printed by Printforce, the Netherlands